ALGEBRA
FOR SCIENCE AND
ENGINEERING STUDENTS

ALGEBRA
FOR SCIENCE AND
ENGINEERING STUDENTS

BY

E. H. LOCKWOOD

Senior Mathematics Master, Felsted School

CAMBRIDGE
AT THE UNIVERSITY PRESS
1940

CAMBRIDGE
UNIVERSITY PRESS

University Printing House, Cambridge CB2 8BS, United Kingdom

Cambridge University Press is part of the University of Cambridge.

It furthers the University's mission by disseminating knowledge in the pursuit of education, learning and research at the highest international levels of excellence.

www.cambridge.org
Information on this title: www.cambridge.org/9781316612736

© Cambridge University Press 1940

First published 1940
First paperback edition 2016

A catalogue record for this publication is available from the British Library

ISBN 978-1-316-61273-6 Paperback

PREFACE

The essential feature of this book is its briefness. Boys studying science require some knowledge of 'Higher Algebra', but unless they are of unusual ability they have not time to work through one of the larger treatises. If their subject is biology, it is usually found that they have not time for mathematics at all, a state of affairs particularly regrettable at a time when biology is becoming every day more mathematical.

A second feature is that probability and statistics are brought within reach of the ordinary sixth form science boy. It is naturally impossible to include more than an introduction to these difficult subjects, but it is quite wrong that science students, particularly biology students, should go to their universities without having been made aware of even the existence of a side of mathematics whose importance is becoming more and more apparent.

Lastly, some attempt has been made to indicate by means of examples the usefulness to scientists of some of the ideas explained in the text. These isolated 'applications' do not of course represent completely the reason why the scientist must study mathematics—for he is dealing all the time in relations and dependences, and mathematics is the study of these in the abstract—but they are put in to help bridge the gap in the student's mind between two seemingly disconnected parts of his work.

I am greatly indebted to Mr D. McGregor, without whose help and encouragement the book would hardly have been completed; to Mr K. S. Snell, who has been kind enough to read the manuscript and proofs; to these and other critics for various suggestions; and to the Oxford and Cambridge Schools Examination Board for permission to reprint questions from examination papers.

E. H. L.

December, 1939

CONTENTS

V. Permutations and combinations

VI. The binomial theorem (for a positive integral index)

VII. Probability

VIII. Finite series

I. INDICES, LOGARITHMS AND SURDS

1.1. The index laws.

If n is a positive integer, a^n means $aaa\ldots$ (n factors).

If m, n are positive integers, it is easy to see that

$$a^m \times a^n = a^{m+n};$$
$$a^m \div a^n = a^{m-n} \text{ (provided that } m > n\text{)};$$
$$(a^m)^n = a^{mn};$$
$$\sqrt[n]{(a^m)} = a^{m/n} \text{ (provided that } m \text{ is a multiple of } n\text{)}.$$

These are the 'index laws'.

Expressions such as a^0, a^{-n}, $a^{m/n}$ (where m is not a multiple of n) are given such meanings as fit in with the index laws:

$$a^0 = 1 \text{ (for } a^0 = a^{3-3} = a^3 \div a^3 = 1\text{)};$$
$$a^{-n} = 1/a^n \text{ (for } a^{-n} = a^{0-n} = a^0 \div a^n = 1/a^n\text{)};$$
$$a^{m/n} = \sqrt[n]{(a^m)} \text{ [for } (a^{m/n})^n = a^m\text{]}.$$

1.2. Logarithmic form of the index laws.

If $x = 10^a$, a is called the 'logarithm of x to the base 10'.

Logarithms are thus indices, and the index laws may be rewritten as follows:

$$\log xy = \log x + \log y;$$
$$\log(x/y) = \log x - \log y;$$
$$\log x^n = n \log x;$$
$$\log \sqrt[n]{x} = \frac{1}{n} \log x.$$

Proofs: We shall assume that the base of the logarithms is 10, but the proofs are general.

Let $\qquad\qquad\qquad \log x = a, \quad \log y = b.$

Then $\qquad\qquad\qquad x = 10^a, \quad y = 10^b$

(for when we say, for example, that $\log 2 = 0\cdot3010$, we mean that $2 = 10^{0\cdot3010}$).

Therefore $\qquad\qquad xy = 10^a \times 10^b = 10^{a+b}.$

Therefore $\qquad\qquad \log xy = a + b = \log x + \log y.$

And similarly for the others.

1.3. Change of base.

It was assumed, in the above proof, that the base of the logarithms was 10. A similar proof would, of course, hold good if another base were used. The only other base of practical importance is e ($= 2 \cdot 718 \ldots$, the sum of the infinite series $1 + 1/1! + 1/2! + 1/3! + \ldots$), chosen on account of the properties that

$$\frac{d}{dx} e^x = e^x, \text{ whereas } \frac{d}{dx} 10^x \simeq 2 \cdot 302 \times 10^x,$$

and

$$\frac{d}{dx} \log_e x = 1/x, \text{ whereas } \frac{d}{dx} \log_{10} x \simeq 0 \cdot 4343 \times 1/x.$$

A change of base may be easily effected by the following rule:

$$\log_b x = \log x / \log b,$$

the logarithms on the right-hand side being to any base.

Proof: Let
$$\log_b x = y.$$

Then
$$x = b^y.$$

Taking logarithms to 10 or any other base,
$$\log x = y \log b.$$

Therefore
$$y = \log x / \log b. \qquad \text{Q.E.D.}$$

If the base chosen for the right-hand side is x, we have the special case

$$\log_b x = 1/\log_x b.$$

Thus, for example, $\log_e 10$ is the reciprocal of $\log_{10} e$.

1.31. It should be noted that $\log_e x$ and e^x are 'inverse functions' (i.e. if x and y are interchanged in the equation $y = \log_e x$, we obtain $x = \log_e y$, which is equivalent to $y = e^x$). This explains the relationship between the graphs of the two functions.

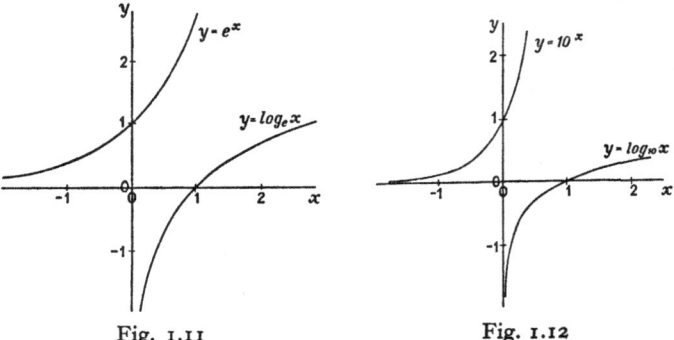

Fig. 1.11   Fig. 1.12

1.4. Surds.

Such numbers as $\sqrt{a}$, $1/\sqrt[3]{a}$, $(\sqrt[3]{a})^2$ are usually dealt with more conveniently by means of indices, in the forms $a^{\frac{1}{2}}$, $a^{-\frac{1}{3}}$, $a^{\frac{2}{3}}$ respectively, but it is also worth noting that a denominator in surd form can be 'rationalised' if we multiply numerator and denominator by a suitable factor.

For example,

$1/\sqrt{2} = \sqrt{2}/2$ (multiplying numerator and denominator by $\sqrt{2}$);

or again, $\qquad 1/(\sqrt{x}+\sqrt{a}) = (\sqrt{x}-\sqrt{a})/(x-a)$

(multiplying numerator and denominator by $\sqrt{x}-\sqrt{a}$).

1.5. Equations involving surds.

In such an equation as

$$\sqrt{(2x+4)}+\sqrt{(x-5)} = 5,$$

it is usual to regard the $\sqrt{}$ sign as meaning the positive square root. The solution of such an equation involves at some point the squaring of both sides, and as this introduces extra solutions, it is essential to test any results so obtained by substituting in the original equation.

Thus the above equation may be solved as follows:

$$\sqrt{(2x+4)} = 5-\sqrt{(x-5)}.$$

Squaring, $\qquad 2x+4 = 25-10\sqrt{(x-5)}+x-5.$

Therefore $\qquad 10\sqrt{(x-5)} = 16-x.$

Squaring again, $\qquad 100x-500 = 256-32x+x^2.$

Therefore $\qquad x^2-132x+756 = 0.$

Therefore $\qquad (x-6)(x-126) = 0,$

and $\qquad x = 6 \quad \text{or} \quad 126.$

If $x = 6$, $\qquad \sqrt{(2x+4)}+\sqrt{(x-5)} = \sqrt{16}+\sqrt{1} = 5.$

If $x = 126$, $\qquad \sqrt{(2x+4)}+\sqrt{(x-5)} = \sqrt{256}+\sqrt{121} = 27.$

The solution $x = 126$ must therefore be rejected. It is evidently a solution, not of the given equation, but of

$$\sqrt{(2x+4)}-\sqrt{(x-5)} = 5.$$

EXAMPLES I

1. Express in index notation:

(1) $\sqrt[3]{a^2}$, (2) $1/\sqrt{(x-a)}$, (3) $\sqrt{(a^7/a^3)}$,

(4) $\sqrt{a^6}/\sqrt[3]{a^{27}}$, (5) $(a^2)^5 \div (a^3 \times a^7)$, (6) $1/(a^{-\frac{2}{3}})^{\frac{3}{4}}$.

2. Simplify:

(1) $x^m \div \sqrt{x^n}$, (2) $(x^2-2ax+a^2)^m \div (x-a)^n$, (3) $\sqrt[3]{x^{2n}} \times \sqrt{x^{3n}}$,

(4) $(x^{m/n})^{n/m}$, (5) $\sqrt[n]{x^{m+n}} \div \sqrt[n]{x^{2m}}$, (6) $1/x^{-n}$.

3. Find x from each of the following equations, working always in indices:

(1) $x^2 = a^{\frac{4}{3}}$, (2) $x^{\frac{2}{3}} = a^{\frac{4}{3}}$, (3) $x^{-2} = a^{\frac{4}{3}}$,

(4) $a/x = \sqrt{x}$, (5) $(a/x)^{\frac{1}{2}} = (x/a)^{\frac{1}{3}}$, (6) $a^3 x^{-n} = ax$.

4. Multiply $x^{\frac{2}{3}}+2+3x^{-\frac{2}{3}}$ by $x^{\frac{2}{3}}-2+3x^{-\frac{2}{3}}$, and find the value of the product when $x = 8$.

5. If $v = a^3$ and $a^2b = 1$, obtain an expression (not containing a) for b, in the form v^n, giving the value of n.

6. Find the value of x if $2 \times 4^{x-1} = 8^{-x}$.

7. If $b = 100a^{-\frac{2}{3}}$, find the value of a when $b = 1$, and the value of b when $a = 8$. Find also, by means of logarithms, the value of b when $a = 0.8$.

8. If $a = b^{\frac{2}{3}}$ and $b = (ac)^{-\frac{1}{2}}$, express c as a power of a, not containing b.

9. Express each of the following as a single logarithm:

(1) $\log(a+x) - \log(a-x)$, (2) $5\log(a+x)$,

(3) $-\frac{1}{5}\log(a+x)$, (4) $m\log x^2 + n\log x^3$,

(5) $m\log(a+x) - n\log(a-x)$, (6) $-k\log y$.

10. Find x from each of the following equations:

(1) $\log x + \log a = \log b$, (2) $n\log x = \log y$,

(3) $\log x = \log a + n\log b$, (4) $m\log x = n\log a$.

11. Evaluate (1) $\log_2 5$, (2) $\log_7 10$.

12. Prove (1) that $\log_a x - \log_a y = \log_a x/y$, (2) that $\log_a x^n = n\log_a x$.

13. Prove that $\log_e x = \log_e 10 \log_{10} x$.

14. Prove in the manner of §1.3 that $\log_a x = \log_a b . \log_b x$.

15. Given that $\log_{10} e = 0\cdot4343$, evaluate:

(1) $\log_e 10$,   (2) $\log_e 8$,   (3) $\log_e 0\cdot8$.

16. Find, to 3 places of decimals, the value of x which satisfies the equation $\log_x 10 = -4$.

17. Rationalise the denominators of:

(1) $1/\sqrt{a}$,   (2) $\sqrt{(x+a)}/\sqrt{(x-a)}$,   (3) $1/(\sqrt{x}-\sqrt{a})$.

18. Given that $\sqrt{2} \simeq 1\cdot414$, evaluate:

(1) $1/\sqrt{2}$,   (2) $1/(\sqrt{2}+1)$,   (3) $(\sqrt{2}+1)/(\sqrt{2}-1)$.

19. Solve the following equations:

(1) $\sqrt{(x+2)} = \sqrt{(2x-3)}-1$,

(2) $\sqrt{(x+7)}+\sqrt{(3x+3)} = 6$,

(3) $\sqrt{(x+1)}+\sqrt{(2x)} = \sqrt{(6x+1)}$.

20. If $y = \dfrac{e^x - e^{-x}}{e^x + e^{-x}}$, express x in terms of y.

21. Given that the differential coefficient of x^n is nx^{n-1} for all values of n, write down the differential coefficients of

(1) $1/x^2$,   (2) $\sqrt[3]{x}$,   (3) $1/\sqrt{x}$.

22. If air is expanded at constant temperature, $pv = $ const., where p, v are the pressure and volume; if it is expanded adiabatically, i.e. without gain or loss of heat, $pv^{1\cdot4} = $ const. If a given quantity of air is at pressure p_0 and is expanded from volume v_0 to v, find the ratio of the final pressure when the expansion is done at constant temperature to the final pressure when it is done adiabatically.

23. If the growth of a part of a living organism is represented by the formula $y = a \cdot b^{c^t}$, where b and c are fractions, and t is the time, measured from the first observation, consider the value of y when $t = 0$ and when t is very large, and hence interpret the quantity a and the fraction b.

24. It is found by plotting $\log y$ against $\log x$ that the variables x and y are related by the equation

$$\log y = 1\cdot301 + 3\log x.$$

Express y in terms of x.

25. Some air is expanded and the temperature and volume are found to follow the law $\log t + 0\cdot4\log v = c$, where c is a constant. Obtain an equivalent relation not involving logarithms and using the fact that if p is the pressure, pv/t is constant, prove that the expansion is adiabatic (i.e. that $pv^{1\cdot4}$ is constant).

26. Coulomb's Law for the force F between two electric charges e, e' at a distance r apart in a vacuum is $F = ee'/r^2$. The 'dimensions' of F and r are respectively $[MLT^{-2}]$ and $[L]$. Show that the unit of electric charge must be regarded as having dimensions $[M^{\frac{1}{2}}L^{\frac{3}{2}}T^{-1}]$.

The force between two magnetic poles m, m' is likewise given by $F = mm'/r^2$, and magnetic intensity H by $H = m/r^2$. Find the dimensions of H.

An electric current is measured by q/t (where q is charge and t is time) and also by $rH \tan \theta/2n\pi$, where r is a length, H a magnetic intensity and n, π, $\tan \theta$ are of no dimensions. Show that on these assumptions q is of dimensions $[M^{\frac{1}{2}}L^{\frac{1}{2}}]$. What are the dimensions of the ratio of these two units of charge?

27. If the pressure p of the atmosphere at height z is given by $p = p_0 e^{-z/H}$, where p_0 and H are constants, prove that the difference in height of two stations (heights z_1, z_2) at which the pressures are p_1, p_2 respectively is given by
$$z_2 - z_1 = 2 \cdot 3026 H \log_{10}(p_1/p_2).$$

28. If l_1, l_2 measure the 'brightness' of two stars whose 'magnitudes' are m_1, m_2, then $l_1/l_2 = k^{m_2 - m_1}$, where $\log k = 0 \cdot 4$. If one star is 100 times as 'bright' as another, by how many units do their 'magnitudes' differ?

29. The gradient of the chord joining two points $(a, \sqrt{a})$ and $(b, \sqrt{b})$ on the curve $y = \sqrt{x}$ is $(\sqrt{b} - \sqrt{a})/(b - a)$. If $b \to a$, the chord becomes the tangent at $(a, \sqrt{a})$. What is the gradient of this tangent?

30. If the sizes x and y of two parts of an organism are represented by the formulae $x = A \cdot B^{kt}$ and $y = a \cdot b^{kt}$, where t represents the time and the other letters are constants, prove that $y = Cx^n$, where $n = \log b/\log B$ and $C = a/A^n$.

II. VARIATION

2.1. It is assumed that the reader has some knowledge of variation, obtained from work on graphs. The present chapter is a summary of some of the more important types of variation, arranged particularly with a view to showing how a formula may be obtained to fit a given set of values of two variables.

2.11. The variable plotted horizontally (x in the examples given) is called the 'independent variable'. We eventually express the other one (called the 'dependent variable') in terms of it (i.e. y in terms of x).

2.12. The term 'gradient', applied to a straight line graph, means

$$\frac{\text{increase in } y}{\text{corresponding increase in } x},$$

the increases being measured according to the scales chosen for y and x respectively. The gradient is taken as $+$ if y increases as x increases, and $-$ if y decreases as x increases.

2.13. The table of values, the graph and the equation all give the same information, but in different ways and with different degrees of completeness.

2.2. Direct proportion.

$$y = ax.$$

x	2	3	4	6	-2
y	6	9	12	18	-6
y/x	3	3	3	3	3

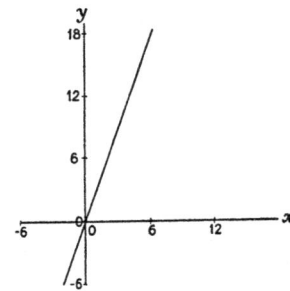

Fig. 2.1

Characteristics: As x increases, y increases in the same ratio.

As x increases by equal steps, so does y.

When $x = 0$, $y = 0$.

Arithmetical test: The quotient y/x or x/y is constant.

Graphical test: The graph is a straight line through the origin.

To find the equation from the graph:

If the gradient is m, $y/x = m$, or $y = mx$.

In the example given, $y = 3x$.

2.3. The linear law. $y = ax + b$.

x	-10	0	10	20
y	0	3	6	9

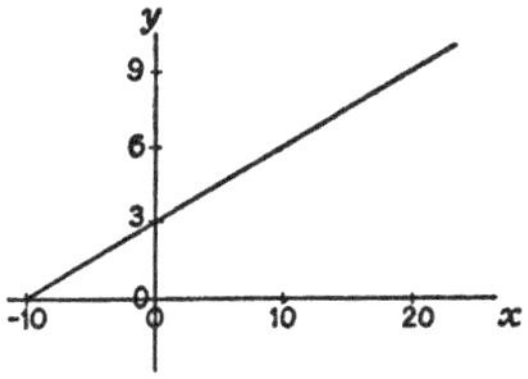

Fig. 2.2

Characteristics: As x increases by equal steps, so does y.

Arithmetical test: Differences between values of y are proportional to those between the corresponding values of x.

Graphical test: The graph is a straight line.

To find the equation from the graph:

Method 1. If the graph is seen to cut the y-axis at the point $y = c$, and the gradient is m: then $y - c$ is the height of any point (x, y) on the line above the point $(0, c)$, and $(y - c)/x = m$. Therefore $y = mx + c$.

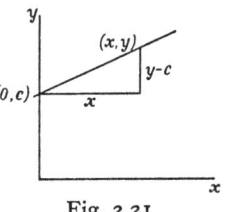

Fig. 2.21

In the example given, gradient $= \frac{3}{10}$ and the starting-point is $(0, 3)$. Therefore $y = \frac{3}{10}x + 3$.

Method 2. This method is useful when the origin, $(0, 0)$, is off the paper.

If the graph is seen to pass through a point (x_1, y_1) and to have gradient m: then $y - y_1$ and $x - x_1$ are the distances of any point (x, y) above and to the right of (x_1, y_1). Hence for any point on the line, $(y - y_1)/(x - x_1) = m$, or $y - y_1 = m(x - x_1)$.

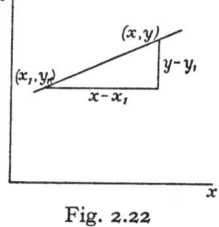

Fig. 2.22

In the example given, the gradient is $\frac{3}{10}$, and if we take $(10, 6)$ as the starting-point, $(y - 6)/(x - 10) = \frac{3}{10}$. Therefore $y = \frac{3}{10}x + 3$.

Method 3. An alternative method, which some find easier, is to select two points on the line, e.g. $(0, 3)$ and $(20, 9)$, and substitute in the equation $y = ax + b$.

Thus $3 = a.0 + b$ and $9 = a.20 + b$, whence $b = 3$ and $a = \frac{3}{10}$.

The equation $y = ax + b$ may then be rewritten as $y = \frac{3}{10}x + 3$.

2.4. Inverse proportion. $y = a/x$.

x	2	3	4	6	-2
y	6	4	3	2	-6
xy	12	12	12	12	12

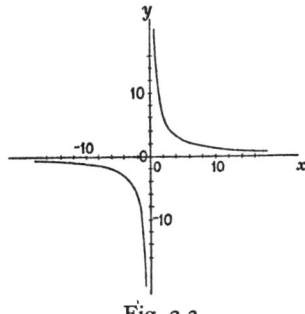

Fig. 2.3

This curve is called a 'hyperbola'.

Characteristics: As x increases, y decreases in inverse ratio.

As $x \to \infty$, $y \to 0$.

As $x \to 0$, $y \to \infty$.

If x changes sign, so does y. (The graph has central symmetry about the origin.)

Arithmetical test: Product xy = constant.

Graphical test: If y is plotted against $1/x$, the graph is a straight line through the origin.

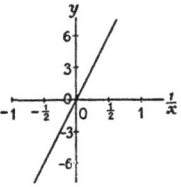

Fig. 2.31

To find the equation from the graph:

If the gradient of the graph in Fig. 2.31 is m, $y \div 1/x = m$,

$$\text{or} \quad xy = m,$$

$$\text{or} \quad y = m/x.$$

In the example given, the gradient is 12. Therefore $y = 12/x$.

2.5. Variation as the square. $y = ax^2$.

x	0	2	3	4	6	-2
y	0	12	27	48	108	12
x^2	0	4	9	16	36	4

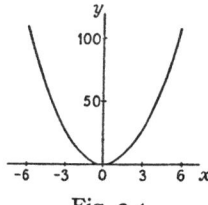

Fig. 2.4

This curve is called a 'parabola'.

Characteristics: As the numerical value of x increases by equal steps, that of y also increases, but by steps of increasing size.

If x changes sign, y is unaltered. (The graph is symmetrical about the y-axis.)

Arithmetical test: The quotient y/x^2 is constant.

Graphical test: If y is plotted against x^2, the graph is a straight line through the origin.

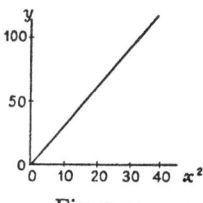

Fig. 2.41

To find the equation from the graph:

If the gradient of the graph in Fig. 2.41 is m, $y/x^2 = m$, therefore $y = mx^2$.

In the example given, the gradient is 3; therefore $y = 3x^2$.

2.6. The inverse square law. $y = a/x^2$.

x	2	3	4	6	-2
y	12	5·33	3	1·33	12
x^2	4	9	16	36	4

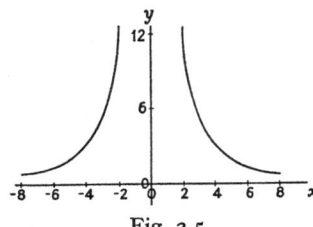

Fig. 2.5

Characteristics: As the numerical value of x increases, that of y decreases.

As $x \to \infty$, $y \to 0$.

As $x \to 0$, $y \to \infty$.

If x changes sign, y is unaltered. (The graph is symmetrical about the y-axis.)

Arithmetical test: The product $x^2 y$ is constant.

Graphical test: If y is plotted against $1/x^2$, the graph is a straight line through the origin.

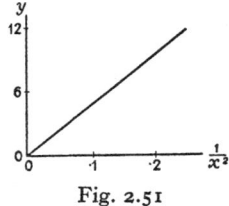

Fig. 2.51

To find the equation from the graph:

If the gradient of the graph in Fig. 2.51 is m, then

$$y \div 1/x^2 = m;$$

therefore

$$x^2 y = m,$$

or

$$y = m/x^2.$$

In the example given, the gradient is 48.

Therefore

$$y = 48/x^2.$$

2.7. y proportional to any power of x.

$$y = ax^n. \tag{1}$$

$$\log y = \log a + n \log x. \tag{2}$$

x	2	3	4	6
y	5·66	10·40	16	29·4

Graphical test: If $\log y$ is plotted against $\log x$, the graph is a straight line.

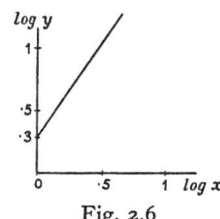

Fig. 2.6

To find the equation from the graph:

Proceeding as in § 2.3, we obtain

$$\log y = 0\cdot3 + 1\cdot5 \log x.$$

By comparison with equation (2) we see that

$$\log a = 0\cdot3 \quad \text{and} \quad n = 1\cdot5.$$

Therefore $a = 2$, approx.

Therefore the equation may be written as

$$y = 2x^{1\cdot5}.$$

2.8. It may also happen that we know the form of the equation connecting two variables and are given only just enough pairs of values to enable us to determine the constants. Thus if $C°$ and $F°$ denote corresponding temperatures on the Centigrade and Fahrenheit scales respectively, we know that the law is a linear one, and we also know the following pairs of values:

C	0	100
F	32	212

Taking the equation as $F = a + bC$,

we obtain by substitution $32 = a + 0$,

$$212 = a + 100b.$$

From these equations, $a = 32$ and $b = \frac{9}{5}$.

Hence the equation may be written as $F = 32 + \frac{9}{5}C$.

2.9. Joint variation.

If $z = axy^2$, a being constant, z is said to 'vary jointly' as x and the square of y. It will be seen that if y is kept constant, z varies as x, and if x is kept constant, z varies as the square of y.

In the study of gases, for example, it is found that the pressure varies inversely as the volume if the temperature is kept constant, and directly as the temperature if the volume is constant. These facts can be combined in the equation $p = kt/v$, or $pv/t = k$, where k is a constant.

EXAMPLES II

1. Find by inspection formulae which express the relationships between the following sets of numbers:

(1)	x	5	6	8	10
	y	8	6·67	5	4
(2)	x	5	6	8	10
	y	8	7	5	3
(3)	x	5	6	8	10
	y	13	16	22	28
(4)	x	5	6	8	10
	y	12·5	18	32	50
(5)	x	5	6	8	10
	y	12·5	21·6	51·2	100
(6)	x	5	6	8	10
	y	12	8·33	4·69	3

2. The effort, P lb., required to raise a load, W tons, by means of a certain machine, is known to be given by a formula of the type

$$P = aW + b.$$

If 32 lb. raises half a ton, and 50 lb. raises a ton, find the exact formula for P in terms of W.

3. Two variables x and y are known to be connected by an equation of the form $y = ax^2 + b$, where a and b are constants. If $y = 77$ when $x = 4$, and $y = 242$ when $x = 7$, find a and b, and the value of x for which $y = 42$.

4. If y varies inversely as the square of x, and $y = 8$ when $x = 5$, give a formula connecting y and x, and find y when $x = 4$.

5. If z varies jointly as y and as the square of x, and $z = 10$ when $y = 5$ and $x = 3$, find the value of z when $x = 5$ and $y = 3$.

6. If z varies as the square of x and inversely as y, and $z = 8$ when $x = 2$ and $y = 5$, find the value of z when $x = 1$ and $y = 4$.

7. The resistance (R) to the motion of a car is given by $a + x$, where a is constant and x varies as the square of the velocity (V). When $V = 15$ miles per hour, $R = 40$ lb. wt., and when $V = 25$ miles per hour, $R = 60$ lb. wt. Find V when $R = 90$ lb. wt.

8. If z varies directly as x/y^2 and y varies inversely as x, and $z = \frac{1}{3}$, $x = 2$, $y = \frac{1}{4}$ are simultaneous values, express (1) y in terms of x, (2) z in terms of x. If the value of x is increased by 10 per cent, find the corresponding percentage of increase in the value of z.

9. The illumination (I units) of a small screen varies directly as the strength (s units) of the source of light, and inversely as the square of the distance (d feet) from the screen. If unit illumination is given by a source of light of unit strength distant 3 feet from the screen, obtain a formula giving I in terms of s and d.

10. The total time taken by a council meeting consists of a fixed number of minutes (for business which does not provoke discussion) added to the time spent in discussion, which varies as the square of the number of members present. With 14 members present the meeting takes 72 minutes, and with 20 members present it takes 123 minutes. How long would the meeting take with 18 members present?

11. Obtain a formula from each of the following graphs:

(1)

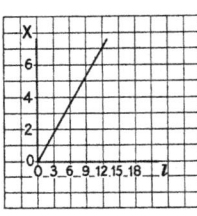

Fig. 2.71

(2)

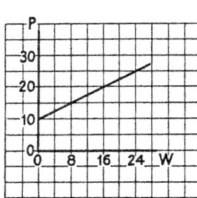

Fig. 2.72

(3)

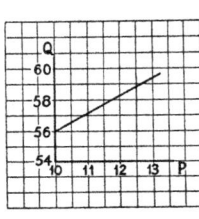

Fig. 2.73

(4)

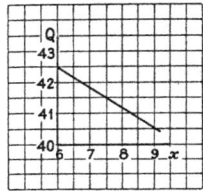

Fig. 2.74

(5)

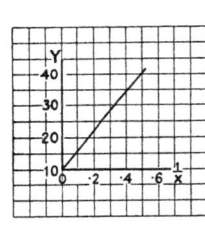

Fig. 2.75

(6)

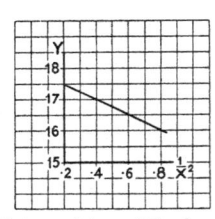

Obtain a relation of the form
$$X^2 Y = a + b X^2$$
Fig. 2.76

(7)

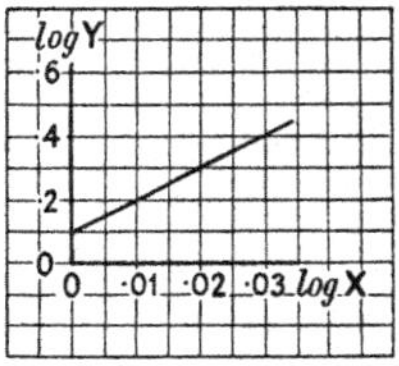

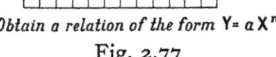

Obtain a relation of the form $Y = a X^n$
Fig. 2.77

(8)

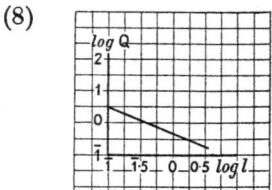

Obtain a relation of the form $Q = a l^n$
Fig. 2.78

12. The following table gives observed values of y corresponding to values of x:

x	5	10	15	20	30
y	4	3	2·7	2·5	2·3

Plot values of xy against values of x. From the resulting graph deduce a simple relation between xy and x, and express y in terms of x.

13. The following table shows corresponding values of x and y:

x	2	4	6·3	10
y	3·2	5	6·8	9·3

Also, the relation between x and y is known to be of the form $y = cx^n$. Plot values of $\log y$ against the corresponding values of $\log x$, and deduce from the resulting graph the values of c and n.

14. It is known that V and H are connected by a relation of the form $V = aH^n$, and corresponding values of V and H are given by the table:

H	2	4	7	10
V	3·25	4·60	6·09	7·27

Plot values of $\log$ V against corresponding values of $\log$ H and from the graph obtained determine approximate values of a and n.

15. The following tables* give the 'life-table death-rate' for towns of different population density (a) in the decade 1861–1870, (b) in 1891–1900:

1861–1870		1891–1900	
δ	D	δ	D
Density per sq. mile	Life-table death-rate	Density per sq. mile	Life-table death-rate
166	19·90	136	17·38
186	21·07	181	18·62
379	23·47	457	20·24
1,718	26·09	1,705	22·71
4,499	28·54	4,884	24·72
12,357	32·67	7,480	27·58
65,823	37·17	55,563	33·25

Verify Farr's Law, that $D = c\delta^m$, where c and m are constants, finding the approximate values of c and m for each of the two decades.

* Reprinted, by permission of the Controller of H.M. Stationery Office, from Dr John Brownlee's *The Use of Death-rates as a Measure of Hygienic Conditions*.

III. THE THEORY OF QUADRATIC EQUATIONS

3.1. Real, equal and imaginary roots.

If
$$ax^2 + bx + c = 0, \qquad (1)$$

$$x^2 + \frac{b}{a}x = -\frac{c}{a},$$

$$x^2 + \frac{b}{a}x + \left(\frac{b}{2a}\right)^2 = \frac{b^2}{4a^2} - \frac{c}{a}$$

$$= \frac{b^2 - 4ac}{4a^2}.$$

Therefore
$$x + \frac{b}{2a} = \frac{\pm\sqrt{(b^2 - 4ac)}}{2a},$$

and
$$x = \frac{-b + \sqrt{(b^2 - 4ac)}}{2a} \quad \text{or} \quad \frac{-b - \sqrt{(b^2 - 4ac)}}{2a}. \qquad (2)$$

If $b^2 - 4ac > 0$, there are two real roots;

if $b^2 - 4ac = 0$, the two roots coincide;

if $b^2 - 4ac < 0$, the roots are said to be 'imaginary'.

3.2. Sum and product of the roots.

If the roots are called α and β, it can be seen by actual addition and multiplication of the values given above that

$$\alpha + \beta = -\frac{b}{a} \quad \text{and} \quad \alpha\beta = \frac{c}{a}.$$

This can also be proved, more simply, as follows: The equation whose roots are α, β is $(x - \alpha)(x - \beta) = 0$, i.e. $x^2 - (\alpha + \beta)x + \alpha\beta = 0$.

Equation (1) may be written $x^2 + \frac{b}{a}x + \frac{c}{a} = 0$.

Hence, by comparison, $\alpha + \beta = -b/a$, and $\alpha\beta = c/a$.

3.21. Symmetrical functions of the roots.

From these results many other symmetrical combinations of α and β can be expressed in terms of a, b, c, e.g.

$$\alpha^2 + \beta^2 = (\alpha + \beta)^2 - 2\alpha\beta = b^2/a^2 - 2c/a.$$

$$(\alpha - \beta)^2 = (\alpha + \beta)^2 - 4\alpha\beta = b^2/a^2 - 4c/a.$$

$$\alpha^3 + \beta^3 = (\alpha + \beta)^3 - 3\alpha\beta(\alpha + \beta) = -b^3/a^3 + 3cb/a^2.$$

$$1/\alpha + 1/\beta = (\alpha + \beta)/\alpha\beta = -b/c.$$

3.22. Forming equations.

It is also easy to form equations whose roots are symmetrically related to those of a given equation. For example, the equation whose roots are α^2 and β^2 is

$$x^2 - (\alpha^2 + \beta^2)\, x + \alpha^2 \beta^2 = 0,$$

i.e.

$$x^2 - \frac{b^2 - 2ac}{a^2}\, x + \frac{c^2}{a^2} = 0.$$

or

$$a^2 x^2 - (b^2 - 2ac)\, x + c^2 = 0.$$

3.23. Another method of forming an equation whose roots are related to those of a given equation is by changing the variable. Thus the equation whose roots are $1/\alpha$ and $1/\beta$ may be found by changing x into $1/y$. The new equation is then $a/y^2 + b/y + c = 0$, or $a + by + cy^2 = 0$. The example of §3.22 may similarly be treated by the substitution $y = x^2$, i.e. $x = \sqrt{y}$.

3.3. Maximum and minimum values of quadratic expressions.

The graph of a quadratic function is a parabola, similar to the curve shown in §2.5. The axis is vertical and the vertex may be either upwards or downwards. There is thus either a maximum or a minimum value. This may be found by 'completing the square', as follows:

$$\begin{aligned}
ax^2 + bx + c &= a\left(x^2 + \frac{b}{a}x + \frac{c}{a}\right)\\
&= a\left\{x^2 + \frac{b}{a}x + \left(\frac{b}{2a}\right)^2 + \frac{c}{a} - \frac{b^2}{4a^2}\right\}\\
&= a\left\{\left(x + \frac{b}{2a}\right)^2 + \frac{4ac - b^2}{4a^2}\right\}.
\end{aligned}$$

The least value of $\left(x + \dfrac{b}{2a}\right)^2$ is zero; so if a is positive, the least value of $ax^2 + bx + c$ is $\dfrac{4ac - b^2}{4a}$.

If a is negative, this same value is a maximum. The maximum or minimum value occurs when $\left(x + \dfrac{b}{2a}\right)^2$ is zero, i.e. when $x = -\dfrac{b}{2a}$.

3.31. Maximum and minimum values of certain fractions.

The maximum and minimum values of such an expression as

$$(x^2 + 3)/(2x + 2)$$

may be found either by calculus or as follows:

Let $\qquad\qquad (x^2 + 3)/(2x + 2) = y.$

Then $\qquad\qquad x^2 + 3 = 2xy + 2y.$

Arranging this as a quadratic equation in x,

$$x^2 - 2y \cdot x + (3 - 2y) = 0.$$

Using the result of §3.1, the values of x are imaginary

if $\qquad\qquad 4y^2 - 4(3 - 2y) < 0,$

i.e. if $\qquad\qquad y^2 + 2y - 3 < 0,$

i.e. if $\qquad\qquad (y + 3)(y - 1) < 0,$

i.e. if y lies between -3 and 1.

y can therefore take any value outside these limits, that is to say, any value from $-\infty$ to -3 and from $+1$ to $+\infty$. It has in fact a minimum value of $+1$ and a maximum of -3.

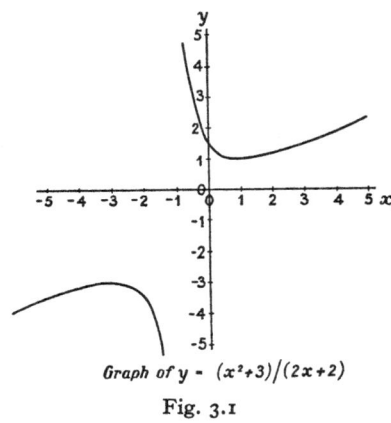

Graph of $y = (x^2 + 3)/(2x + 2)$

Fig. 3.1

EXAMPLES III

1. For what values of a are the roots of the equation $x^2 + 3x + a = 0$ (1) real and different, (2) equal, (3) imaginary?

2. Prove that the roots of $2ax^2 + 2(a + b)x + b = 0$ are always real.

3. If α and β are the roots of $ax^2 + bx + c = 0$, find in terms of a, b, c:

    (1) $\alpha^2\beta + \alpha\beta^2,$ (2) $\alpha^4 + \beta^4,$

    (3) $1/\alpha^2 + 1/\beta^2,$ (4) $\alpha/\beta + \beta/\alpha.$

4. If α and β are the roots of $ax^2 + bx + c = 0$, form the equations whose roots are

    (1) $1/\alpha$ and $1/\beta,$ (2) α^3 and $\beta^3,$

    (3) $\alpha^3\beta$ and $\alpha\beta^3,$ (4) α^2/β and $\beta^2/\alpha.$

5. Find by the method of § 3.2 the sum and the product of the roots of $x^2 - 2px + q = 0$, and use these results to obtain (1) the sum of the squares of the roots, (2) the equation whose roots are the squares of those of the given equation. (Part (2) should be done by two different methods.)

6. If the roots of the quadratic $ax^2 + 2bx + c = 0$ are α and β, express $\alpha^6 + \beta^6$ in terms of a, b, c. Form the equation whose roots are $\alpha + 1/\beta$ and $\beta + 1/\alpha$.

7. If α, β are the roots of the equation $x^2 - x + 5 = 0$, find the equation whose roots are $\alpha^2 + 1/\beta^2$ and $\beta^2 + 1/\alpha^2$.

8. Find the values of k for which the quadratic equation
$$k(x^2 - 10x - 2) + (2x^2 + 1) = 0$$
has equal roots.

9. If α is a root of the equation $x^2 + 7x + 3 = 0$, it follows that
$$\alpha^2 + 7\alpha + 3 = 0.$$
Hence prove that if β is the other root, $(\alpha^2 + \beta^2) + 7(\alpha + \beta) + 6 = 0$. Prove also that $(\alpha^3 + \beta^3) + 7(\alpha^2 + \beta^2) + 3(\alpha + \beta) = 0$.

10. If α, β are the roots of the equation $x^2 - 5x + 7 = 0$, prove that $(\alpha^4 + \beta^4) - 5(\alpha^3 + \beta^3) + 7(\alpha^2 + \beta^2) = 0$. Hence prove that $\alpha^4 + \beta^4 = 23$.

11. Find the least value of the expression $2x^2 - 3x + 5$, stating also the corresponding value of x.

12. Find the greatest value of $2 - 4x - x^2$, stating also the corresponding value of x.

13. Show that the expression $(2x^2 - 3x - 2)/x^2$ can never exceed the value $3\frac{1}{8}$.

14. Show that $(3x - 2x^2 - 2)/x$ can never lie between 7 and -1.

15. Show that the maximum value of $(6x - 2x^2 + 2)/(x^2 + 1)$ is $+\sqrt{13}$ and the minimum $-\sqrt{13}$.

16. Prove that, if a and b have opposite signs, the expression $ax + b/x$ can be made to take any real value by giving a suitable real value to x; but that, if a and b have the same sign, and x is real, the expression cannot take any value lying between $\pm 2\sqrt{(ab)}$. Illustrate these results by indicating the general shapes of the graphs of $ax + b/x$ and $ax - b/x$, when a and b are both positive.

17. Show that if x_1 and x_2 are the roots of
$$ax^2 + bx + c + \lambda(a'x^2 + b'x + c') = 0,$$
then $x_1 x_2(ab' - a'b) - (x_1 + x_2)(ca' - c'a) + bc' - b'c = 0.$

18. If the roots of the quadratic $ax^2 + 2bx + c = 0$ are α and β, obtain in terms of α and β the roots of the equation
$$ax^2 + 2bx + c = 8(b^2 - ac)/a.$$

19. By solving the equations simultaneously, show that the straight line $y = x + c$ cuts, touches or misses the parabola $y = x^2 + 5$ according as c is greater than, equal to or less than $19/4$.

20. A ball thrown vertically upwards from the ground with velocity u passes a point at a height h at two times given by $h = ut - \frac{1}{2}gt^2$, where g is a constant. Show that the two values of t coincide if $u^2 = 2gh$, and interpret the result.

21. A stone is thrown with velocity V to strike an object on the same level as the point of projection and at a horizontal distance R. There are usually two possible angles of projection, given by the equation
$$0 = R\tan\alpha - \tfrac{1}{2}gR^2(1 + \tan^2\alpha)/V^2.$$
By arranging this as a quadratic in $\tan\alpha$ and considering whether the values of $\tan\alpha$ are real or imaginary, prove that the maximum horizontal range is V^2/g.

22. The period of oscillation of a compound pendulum is
$$2\pi\sqrt{\{(k^2 + h^2)/gh\}},$$
where h is the distance of the pivot from the centre of gravity and k, g are constants. Let $y = (k^2 + h^2)/h$, and by considering this as an equation in h, show that the minimum time of oscillation, for different positions of the pivot, is $2\pi\sqrt{(2k/g)}$.

23. The following is a graphical method for solving the equation
$$x^2 - px + q = 0.$$
Plot the points $(0, 1)$, (p, q) and draw a circle on the line joining them as diameter. Let this circle cut the x-axis at the points $(\alpha, 0)$, $(\beta, 0)$. By showing that $\alpha + \beta = p$ and $\alpha\beta = q$, prove that α and β are the roots of the equation.

24. If α, β, γ are the roots of $ax^3 + bx^2 + cx + d = 0$, find $\alpha + \beta + \gamma$, $\beta\gamma + \gamma\alpha + \alpha\beta$, and $\alpha\beta\gamma$ in terms of a, b, c, d. Hence find $\alpha^2 + \beta^2 + \gamma^2$ and $\beta^2\gamma^2 + \gamma^2\alpha^2 + \alpha^2\beta^2$.

IV. FACTORS AND SOME DEVELOPMENTS

4.1. Test for factors.

In dealing with an equation $ax^2 + bx + c = 0$, we know that if $x = \alpha$ is a root, then $x - \alpha$ must be a factor of the L.H.S. (left-hand side). Conversely, if $x - \alpha$ is a factor of a given expression, then the substitution $x = \alpha$ will make the expression 'vanish' (become equal to zero).

Proof: Suppose that the expression $\mathsf{E} = (x - \alpha)(\dots)$.

If $x = \alpha$, the R.H.S. vanishes. So, therefore, does E.   Q.E.D.

This provides a convenient way of testing for factors. E.g. if we want to know whether $x - 3$ is a factor of $x^3 - 7x - 6$, we put $x = 3$. The result is $27 - 21 - 6$, which $= 0$. Therefore $x - 3$ is a factor. If we put $x = -4$, the result is $-64 + 28 - 6$, which $= -42$. So $x + 4$ is not a factor.

4.11. Remainder theorem.

In the example just given, -42 is the remainder obtained on dividing $x^3 - 7x - 6$ by $x + 4$. The remainder on dividing $f(x)$ by $x - \alpha$ is $f(\alpha)$.

Proof: Let $f(x) = (x - \alpha)\mathsf{Q} + \mathsf{R}$, where Q and R stand for 'quotient' and 'remainder'. Put $x = \alpha$. Then $f(\alpha) = \mathsf{R}$.

4.12. Cubic equations.

It is sometimes possible to solve a cubic equation as follows:
$$x^3 + 5x - 18 = 0.$$
Put $x = 1$. L.H.S. $= 1 + 5 - 18 = -12$. $\therefore x = 1$ is not a solution.
Put $x = -1$. L.H.S. $= -1 - 5 - 18 = -24$. $\therefore x = -1$ is not a solution.
Put $x = 2$. L.H.S. $= 8 + 10 - 18 = 0$. $\therefore x = 2$ is a solution.

$$\therefore \ x - 2 \text{ is a factor of the L.H.S.}$$

By division, the other factor is $x^2 + 2x + 9$.

$\therefore$ Either $x = 2$ or $x^2 + 2x + 9 = 0$.

$\therefore$ Either $x = 2$ or $x = \{-2 \pm \sqrt{(4 - 36)}\}/2$.

$\therefore \ x = 2$ is the only real solution.

4.13. Factors of $a^3 + b^3$ and $a^3 - b^3$.

If $a = b$, then $a^3 - b^3 = 0$. So $a - b$ must be a factor of $a^3 - b^3$.

Similarly, if $a = -b$, $a^3 + b^3 = 0$. So $a + b$ is a factor of $a^3 + b^3$.

By division, we find that $a^3 + b^3 = (a + b)(a^2 - ab + b^2)$
and that $a^3 - b^3 = (a - b)(a^2 + ab + b^2)$.

These two results should be remembered.

4.2. Symmetry.

$a^3 + b^3$ is said to be 'symmetrical in a and b', meaning that if a and b were interchanged the expression would not be altered. The factors of $a^3 + b^3$ also form an expression symmetrical in a and b. This sometimes provides a useful check. E.g. the statement

$$a^4 + a^2 b^2 + b^4 = (a^2 + ab + b^2)(a^2 - ab - b^2)$$

is obviously wrong, because the L.H.S. is symmetrical in a and b, but not the R.H.S.

4.21. Cyclic symmetry.

The expression $a^3(b-c) + b^3(c-a) + c^3(a-b)$ is said to have 'cyclic symmetry', meaning that if a, b, c are replaced respectively by b, c, a, or by c, a, b, the expression is unaltered. It is easy to see that $b - c$ is a factor of this expression, because if $b = c$, the expression vanishes. In view of the cyclic symmetry, it is clear that $c - a$ and $a - b$ must also be factors. But the expression is of fourth degree, so these three factors are not the only ones. There must be one more, of first degree, and symmetrical in a, b, c. The only possibility is $a + b + c$.

Therefore

$$a^3(b-c) + b^3(c-a) + c^3(a-b) = k(b-c)(c-a)(a-b)(a+b+c),$$

where k is a numerical factor. k may be determined by considering a particular term, e.g. the term in $a^3 b$. The coefficient of $a^3 b$ in the L.H.S. is 1. In the R.H.S. the only way in which $a^3 b$ can be formed is from the b in the first bracket and the a in each of the others. So the coefficient of $a^3 b$ in the R.H.S. is $-k$. Therefore $-k = 1$, and $k = -1$.

4.3. Σ notation.

The expression factorised in the preceding paragraph may be written $\Sigma a^3(b-c)$. The Greek letter Σ (pronounced 'sigma') is used to indicate the sum of the terms obtained by changing the letters in cyclic order.

Σ may also be used to indicate the sum of a series. E.g.

$$1^2 + 2^2 + 3^2 + \ldots + n^2$$

may be written $\sum\limits_{r=1}^{n} r^2$.

4.4. Equating of coefficients.

The sign '$\equiv$' is used to indicate that two expressions are 'identically equal', that is, equal for all values of some variable. E.g. $x + 2 = 5$ is an equation, but $x + 2 \equiv 2 + x$ is an identity. For two expressions to be identically equal, the coefficients of each separate power of the variable must be the same in both. This condition is necessary and sufficient. E.g. $1 + 7x + 6x^3$ can be expressed in the form

$$A + Bx + Cx(x+1) + Dx(x+1)(x+2),$$

for it is only necessary to have $1 = A$

$$7 = B + C + 2D$$
$$0 = C + 3D$$
$$6 = D,$$

and values of A, B, C, D can be found to satisfy these four equations.

4.5. Partial fractions.

A fraction whose denominator will factorise can be expressed in 'partial fractions', thus:

$$\frac{3x+2}{(x+5)(x-3)} \equiv \frac{A}{x+5} + \frac{B}{x-3}, \tag{1}$$

where A, B are numbers.

For this identity to be true, we must have:

$$3x + 2 \equiv A(x-3) + B(x+5), \tag{2}$$

and hence, equating coefficients,

$$3 = A + B,$$
$$2 = -3A + 5B.$$

These equations can probably be satisfied, since there are two equations and two unknowns.

The solution is $A = 1\tfrac{3}{8}$, $B = 1\tfrac{1}{8}$.

Therefore $\quad \dfrac{3x+2}{(x+5)(x-3)} \equiv \dfrac{1\tfrac{3}{8}}{x+5} + \dfrac{1\tfrac{1}{8}}{x-3}.$

A quick way of obtaining the values of A and B is to substitute $x = -5$ and $x = 3$ in the identity (2), but if this short cut is employed, it is advisable to make sure first that the number of unknowns is the same as the number of equations that have to be satisfied. This latter requirement explains the necessity for the following forms in more complicated cases:

$$\frac{2}{(x+1)(x+3)^2} \equiv \frac{A}{x+1} + \frac{B}{x+3} + \frac{C}{(x+3)^2},$$

$$\frac{2}{(x+1)(x^2+3)} \equiv \frac{A}{x+1} + \frac{Bx+C}{x^2+3},$$

$$\frac{2}{(x+1)(x^2+3)^2} \equiv \frac{A}{x+1} + \frac{Bx+C}{x^2+3} + \frac{Dx+E}{(x^2+3)^2},$$

$$\frac{x^2+3}{(x+1)(x+3)} \equiv A + \frac{B}{x+1} + \frac{C}{x+3},$$

$$\frac{2x^3+x^2+3}{(x+1)(x+3)} \equiv Ax + B + \frac{C}{x+1} + \frac{D}{x+3}.$$

It will be noticed that in each of the last two examples the value of the first coefficient, A, is obvious by inspection and can be written in at once. For if x^2+3 were divided by $(x+1)(x+3)$, i.e. by x^2+4x+3, the quotient would obviously be 1. Therefore A = 1. In the last example, both A and B could be obtained by long division.

EXAMPLES IV

1. Find the linear (i.e. first degree) factors of $x^3 - 2x^2 - 5x + 6$.

2. Solve the equation $x^3 + 8x + 9 = 0$.

3. Solve the equation $x^3 + 3x^2 - 34x + 48 = 0$.

4. Factorise $x^3 + 3x^2 + 5x + 6$.

5. What is the remainder if $x^3 - 5x$ is divided by $x+4$? Check by division.

6. What is the remainder if $x^4 + a^4$ is divided by $x+a$?

7. Test whether $ax + b$ is a factor of $a^3x^3 + 7a^2bx^2 + 3ab^2x - 3b^3$.

8. State the factors of $1 + a^3x^3$ and of $27x^3 - y^3$.

9. Show that $b - c$ is a factor of $a^2(b-c) + b^2(c-a) + c^2(a-b)$, and find the remaining factors.

10. Factorise $a(b-c)^3 + b(c-a)^3 + c(a-b)^3$.

11. Factorise $a(b+c)^2 + b(c+a)^2 + c(a+b)^2 - 4abc$.

12. Factorise $a^2(b+c)+b^2(c+a)+c^2(a+b)+2abc$.

13. Use Σ notation to abbreviate the expressions mentioned in Exx. 10, 11, 12.

14. Simplify $\Sigma a(b-c)$ and $\Sigma a(b+c)$.

15. State the meaning of $\overset{n}{\underset{r=1}{\Sigma}} r$ and $\overset{n}{\underset{r=1}{\Sigma}} (2r+3)$.

16. Express x^3+1 in the form $x(x-1)(x-2)+\mathsf{A}x(x-1)+\mathsf{B}x+\mathsf{C}$.

17. Show that $px+q$ may be expressed in the form $\mathsf{A}(x-\alpha)+\mathsf{B}(x-\beta)$, where A and B are independent of x. Determine A and B in terms of p, q, α, β.

18. Express n^4 in the form
$$\mathsf{A}n(n+1)(n+2)(n+3)+\mathsf{B}n(n+1)(n+2)+\mathsf{C}n(n+1)+\mathsf{D}n.$$

19. The quadratic expression px^2+qx+r is required to take the values A, B, C when x is equal to a, b, c respectively. Show that, in general, p, q, r can be chosen in such a way that this is so. Observe that the expression
$$\mathsf{A}\frac{(x-b)(x-c)}{(a-b)(a-c)}+\mathsf{B}\frac{(x-c)(x-a)}{(b-c)(b-a)}+\mathsf{C}\frac{(x-a)(x-b)}{(c-a)(c-b)}$$
satisfies the conditions. Hence write down the values of p, q, r. (Use Σ notation.)

20. Express in partial fractions:

(1) $\dfrac{3}{(x+1)(x+4)}$,   (2) $\dfrac{4x+7}{(x+1)(x+3)}$,

(3) $\dfrac{x^2+5x+6}{x^2+6x+5}$,   (4) $\dfrac{3}{x(x-1)^2}$,

(5) $\dfrac{5x+3}{x^3+1}$,   (6) $\dfrac{x^3}{x^3-3x+2}$.

21. Express in partial fractions:

(1) $\dfrac{x}{(x-1)(x-2)^2}$,   (2) $\dfrac{2x^3}{x^3-1}$,   (3) $\dfrac{2x+1}{x^4-1}$.

22. Show that
$$\frac{1}{(x-a)(x-b)}\equiv\frac{\mathsf{A}}{x-a}-\frac{\mathsf{B}}{x-b},$$
where
$$\mathsf{A}=\mathsf{B}=1/(a-b).$$

23. Show that
$$\frac{x^2}{(x-a)(x-b)}\equiv 1+\frac{a^2}{(a-b)(x-a)}+\frac{b^2}{(b-a)(x-b)}.$$

24. Express $\dfrac{x^3+1}{x^2-4}$ in the form $ax+b+\dfrac{c}{x-2}+\dfrac{d}{x+2}$.

25. Express $\dfrac{x^4+1}{x^2+3x-4}$ in partial fractions.

26. Given that the integral of $k/(x+a)$ is $\log(x+a)^k$, find the integral of $(3x+2)/(x^2-9)$.

27. Given that, if x lies between -1 and $+1$,

$$1/(1+x) = 1-x+x^2-x^3+\cdots$$

expand $1/(1+5x+4x^2)$, stating for what values of x the expansion is valid.

28. Prove by the Remainder Theorem that $a+b+c$ is a factor of $a^3+b^3+c^3-3abc$. What conditions must a and b satisfy in order that $x^3+a^3+b^3-3abx$ may be identically equal to x^3-6x+7? Form a quadratic equation whose roots are a^3 and b^3, and hence find a root of the equation $x^3-6x+7=0$.

29. Show that the equation $x^3-3qx+r=0$ can be solved by the method of Ex. 28, provided that r^2-4q^3 is greater than or equal to zero.

(Note. If r^2-4q^3 is positive, there is only one real root. If r^2-4q^3 is negative or zero, there are three real roots, and they may be obtained by substituting $x=k\cos\theta$, where $k=2\sqrt{q}$, and using the identity

$$4\cos^3\theta-3\cos\theta = \cos 3\theta.)$$

V. PERMUTATIONS AND COMBINATIONS

5.1. Terminology.

A selection of three letters of the alphabet, such as

is called a 'combination'.

An arrangement of three letters in order, such as AEV or AVE or ABC, is called a 'permutation'.

It will be observed that the number of permutations is greater than the number of combinations, for each combination of 3 letters can be arranged in 6 different ways, to form 6 different permutations, e.g. AEV, AVE, EAV, EVA, VAE, VEA. By methods to be explained below, it can be shown that a selection, or combination, of 3 different letters of the English alphabet can be chosen in 2600 ways; an arrangement, or permutation, of 3 different letters can be chosen in 15,600 ways.

5.2. Notation.

As it is often convenient to refer to these numbers without working them out, their descriptions as 'the number of combinations that can be chosen from 26 different things, taking 3 at a time' and 'the number of permutations of 26 different things, taking 3 at a time' are abbreviated respectively to $_{26}C_3$ and $_{26}P_3$.

Speaking generally, $_nP_r$ denotes 'the number of permutations from n things, r at a time', and $_nC_r$ 'the number of combinations from n things, r at a time'.

5.3. Evaluation of $_nP_r$.

The first thing may be chosen in n ways; for each of these ways, the second thing may be chosen in $(n-1)$ ways; so, combining these facts, the first two things may be chosen in $n(n-1)$ ways. For each of these ways of choosing two things, the third may be chosen in $(n-2)$ ways, the fourth in $(n-3)$ ways, and so on till we reach the rth, which may be chosen in $\{n-(r-1)\}$ ways, i.e. in $(n-r+1)$ ways.

Therefore $\quad _nP_r = n(n-1)(n-2)\ldots(n-r+1) = n!/(n-r)!,$ \qquad (1)

where $n!$ denotes 'factorial n', i.e. the product $1.2.3\ldots n$.

For example, 3 different letters might be chosen for a car registration mark in $26.25.24$ ways.

5.31. Permutations with repetitions.

If the same letter may be used more than once, the 3 letters for the registration mark can be chosen in $26.26.26$, i.e. 26^3, ways, the argument being similar to that used above.

In general, if repetitions are allowed, r things can be chosen in order from n different kinds of things in n^r ways.

5.32. Permutations of n things, n at a time.

As a special case, $_nP_n = n(n-1)(n-2)\ldots 2.1,$
$$= n!,$$

i.e. n things can be arranged among themselves in $n!$ different orders.

It will be noted that the expression on the extreme right of equation (1) will not reduce to this unless we agree that the symbol $0!$, at present undefined, shall have the value 1. Henceforward, therefore, $0! = 1$.

5.4. Evaluation of $_nC_r$.

We must first observe that $_nC_r$ is always less than $_nP_r$ (unless $r = 1$, when they are equal), because the things forming one combination can be re-arranged among themselves to form a number of permutations. To be exact, r things can be re-arranged among themselves in $r!$ ways (as in § 5.32), so
$$_nP_r = r!\,_nC_r.$$

Therefore $_nC_r = {}_nP_r/r! = \dfrac{n(n-1)\ldots(n-r+1)}{r!} = \dfrac{n!}{(n-r)!\,r!}.$ (2)

For example, we can select 3 cards from a pack in $_{52}C_3$ ways, i.e.
$$\frac{52.51.50}{1.2.3}.$$

It should be noted

(1) that the number of factors is the same in the numerator as in the denominator;

(2) that the expression must 'cancel down' to a whole number;

(3) that as it may be written $\dfrac{52!}{3!\,49!}$, it is equal to $_{52}C_{49}$.

More generally, $_nC_r = {}_nC_{n-r}$, a fact which is obvious enough if we consider that a selection of r things is also a rejection of $(n-r)$ things.

5.41. Special cases of $_nC_r$.

$$_nC_1 = {}_nC_{n-1} = n.$$
$$_nC_0 = {}_nC_n = 1.$$

5.5. To divide n things into groups of p, q, r, . . ., where

$$p + q + r + \ldots = n.$$

(1) If p, q, r. . .are all different:

The first group can be chosen in $_nC_p$ ways, the second in $_{n-p}C_q$ ways, etc.

Therefore the required number is $_nC_p \times _{n-p}C_q \times _{n-p-q}C_r \times \ldots$.

This is equal to

$$\frac{n!}{p!(n-p)!} \times \frac{(n-p)!}{q!(n-p-q)!} \times \ldots.$$

The last factor in the denominator will be $0!$, and the expression is therefore equal to

$$\frac{n!}{p!\,q!\,r!\ldots}.$$

The formula for $_nC_r$ may be regarded as a special case of this result, the n things being divided into 2 groups of n and $n-r$ things respectively.

(2) If two or more groups contain equal numbers of things:

There may or may not be other means of distinguishing the groups. If 8 oarsmen are to be divided into 2 fours, one of which is to be regarded as 'the 1st four', the number of ways of making the selection is $\frac{8!}{4!\,4!}$. But if there is to be no distinction between the 2 fours (if, for example, it is desired that they should be of equal strength, for a race against each other), then we have counted every way of making the selection twice (for it makes no difference whether A, B, C and D row against E, F, G and H or vice versa). The result is therefore $\frac{1}{2} \cdot \frac{8!}{4!\,4!}$.

Similarly, if 3 of the groups are indistinguishable, we have counted each selection $3!$ times, and the correct result is $\frac{n!}{3!\,p!\,q!\,r!\ldots}$. And so on.

5.6. Total number of combinations of n things.

I.e. $\qquad\qquad _nC_0 + _nC_1 + _nC_2 + \ldots + _nC_n.$

This is easily evaluated by considering the n things in turn. The first thing may be either selected or rejected, i.e. it may be dealt with in either of 2 ways. For each of these the second thing may be dealt with in 2 ways, and so on. The result is therefore 2^n. If, however, $_nC_0$ (representing the rejection of all the n things) is not to be included, then the result is $2^n - 1$.

EXAMPLES V

1. The numbers at a telephone exchange run from 100 to 999 inclusive. Find how many there are (1) by subtraction, (2) by considering in how many ways each digit may be chosen.

2. The registration marks of motor vehicles used to consist at one time of one letter followed by 4 numerals, the numerals being any digit from 0 to 9 inclusive, except that they could not all four be 0. How many possible registration marks were there?

3. The registration marks of motor vehicles were recently changed from 2 letters and 4 numerals to 3 letters and 3 numerals. Show that there are approximately 2·6 times as many possible marks in the new series as in the old. (Neglect the proviso that the numerals cannot all be 0.)

4. If everyone surnamed Smith had two Christian names, how many possible different pairs of initials would there be? If none of them could have the same letter for both initials, how many possible pairs would there be then?

5. An elector has to choose 3 out of 10 candidates, writing down their names in order of preference. In how many ways can he vote?

6. A committee of 10 has to appoint 3 of its members as a sub-committee. In how many ways can a choice be made?

7. Ten different books are to be divided between two prize-winners, one of whom is to receive a first prize of 7 books, the other a second prize of 3. In how many ways can the books be divided?

8. A telephone number in the London area consists of 3 letters and 4 digits. In dialling a letter there are 9 holes in which the finger may be placed; for a digit there are 10. How many possible telephone numbers are there on this system? By how many is the possible number reduced if the 4 digits cannot all be zero?

9. In how many ways can the 4 positions in the three-quarter line be filled from the 15 men in a Rugby football team? In how many ways can 8 forwards be chosen from a team of 15, assuming that their positions in the scrum are not being considered?

10. If 7 people wish to play tennis, in how many ways can they be divided into 4, 2 and 1? In how many ways can the 4 be divided to make a 'doubles' match?

11. In how many ways can 22 people be divided into 2 cricket teams, (1) to play as 1st XI and 2nd XI against another club, (2) to play against each other in a 'friendly' game?

12. Three people wish to sit together in a theatre. If a row of 12 empty seats is available, in how many ways can they choose their block of 3 seats? What is the total number of ways in which they can arrange themselves?

13. A committee of 8 decides to send a deputation to see the Home Secretary. In how many ways can the deputation be chosen from the 8 members?

14. A man who has 8 pennies in his pocket decides to contribute to a collection. How many different possible contributions can he make?

15. A person entering for a football pool has to forecast 'win', 'lose' or 'draw' for 16 different matches. In how many ways can he make his choice?

16. A conjuror asks a member of his audience to select 5 cards from a pack of 52. In how many ways can the choice be made?

17. A committee of 8 is to include exactly 3 women. In how many ways can it be chosen from 20 women and 30 men?

18. A person is invited to help himself from a plate of 12 cakes, all different. How many different meals can he make? In how many ways can he choose 3 of the cakes? In how many ways can he choose not more than 3?

19. If there are 15 horses entered for a race, and 1st, 2nd and 3rd places count, how many possible results are there?

20. (1) If, in a family of 7, every member gives every other member a Christmas present, how many presents are given altogether?

(2) If, in a party of 7, each person shakes hands with each other person, how many handshakes are there?

21. In the Morse code, what is the total number of letters that could be represented by not more than 4 dots or dashes?

22. There are 12 horses entered for a race. In how many ways can a backer select 3 of them (1) if he proposes to back all three to win, (2) if he intends to back one to win and the other two for places?

23. There are 3 roads from A to B and 4 from B to C. How many routes are there from A to C, using any of these roads, but without returning to A? If a motorist travels from A to B, then takes a wrong turning and finds himself back at A, afterwards taking the same road to B as before, then on to C, how many possible routes are there that he may have taken?

24. A chessboard is placed with one side north-and-south. A piece is moved from the south-west corner to the north-east corner by 16 moves of one square each. How many different possible routes are there?

25. Ten articles are to be placed in a row, three of them, A, B and C, coming together. Prove that this can be done in 241,920 ways. In how many ways can 9 articles be arranged in a row so that two of them, A and B, do not come together?

26. Ten coloured beads are to be arranged on a table in a circle. In how many ways can this be done, (1) when the beads are all of different colours, (2) when three are of the same colour and the rest different?

27. Show that the number of combinations, $n-2$ at a time, of n things, of which three are alike and the rest all different, is $\frac{1}{2}(n^2-5n+8)$.

28. Show that the number of combinations, $n-3$ at a time, of n things, of which four are alike and the rest different, is $\frac{1}{6}(n-3)(n^2-9n+26)$.

29. A certain organism has the genetic constitution $AaBb$; it can produce gametes, in equal numbers, of the 4 types AB, Ab, aB, ab. These may be represented by the 4 terms of the product $\left(\frac{A+a}{2}\right)\left(\frac{B+b}{2}\right)$. If two such organisms are crossed together, the genetic constitution of the offspring will be obtained by combining any two of these gametes; e.g. AABB is obtained from AB and AB. These constitutions will therefore be represented by the terms of the product $\frac{1}{16}(A+a)^2(B+b)^2$, i.e.

$$\tfrac{1}{16}(A^2+2Aa+a^2)(B^2+2Bb+b^2).$$

Show that the proportions of those containing both A and B, A but not B, B but not A, neither A nor B, will be as $9:3:3:1$.

30. A certain organism has the genetic constitution $AaBbCc$; it can produce gametes, in equal numbers, of types ABC, ABc, Abc, abc, etc. How many such types will there be? They will be represented in fact by the terms of the expansion $\frac{1}{8}(A+a)(B+b)(C+c)$. If 2 such organisms are crossed together, the genetic constitution of the offspring will be obtained by combining any two of these gametes; e.g. $Aa\,BB\,Cc$ is obtained from ABC and aBc or from ABc and aBC. How many such constitutions will there be? These constitutions will be terms of the product

$$\tfrac{1}{64}\,(A^2+2Aa+a^2)\,(B^2+2Bb+b^2)\,(C^2+2Cc+c^2).$$

What proportion of the total will be (1) of type $A^2B^2C^2$, (2) of type $Aa\,Bb\,CC$, (3) what proportion will not contain C, (4) what proportion will contain at least one A, at least one B and at least one C?

(*Note.* This question can be related to *Phaseolus*, the garden bean. A plant with neither A, B nor C is short and bushy, having only 3 internodes; one with A or AA adds about 10 internodes; B and C add one more each; BB and CC two more each. Thus $AA\,BB\,Cc$ and $Aa\,Bb\,CC$ will both be 16 internodes tall, and so on.)

31. Two varieties of barley are crossed, one having the genetic constitution $AA\,BB\,CC\,DD$, the other having none of these characteristics. The offspring have the constitution $Aa\,Bb\,Cc\,Dd$. If these are self-fertilised, what proportion of the resulting generation will contain (1) A, B, C and D, (2) A, B, C but not D, (3) A, B but not C or D, (4) A, but not B, C, or D, (5) none of these characteristics?

32. Using the method of example 29, we see that since organisms of constitution AA produce gametes all of which contain A, but those of constitution Aa produce gametes represented by $(A+a)/2$, we may represent the gametes of any type by changing A^2 into A, Aa into $(A+a)/2$, etc., obtaining for the type A^2Bb, for example, the expression $A\,(B+b)/2$, i.e. $(AB+Ab)/2$. Thus if AA and aa are crossed, the gametes are A and a, and the first generation of offspring are all Aa. If this generation is self-fertilised, the gametes are $(A+a)/2$, and the second generation is represented by $\left(\dfrac{A+a}{2}\right)\left(\dfrac{A+a}{2}\right)$, i.e. $(A^2+2Aa+a^2)/4$. If each type in this generation is self-fertilised, the gametes $A/4$, $(A+a)/4$, $a/4$ are combined with gametes A, $(A+a)/2$, a. Show that the result is to produce the three types in the ratio $3:2:3$, and find the ratio for the next generation after this.

33. If each type of the generation obtained in Ex. 29, namely

$$\tfrac{1}{16}\,(A^2+2Aa+a^2)\,(B^2+2Bb+b^2),$$

is self-fertilised, find by the method of Ex. 32 the proportions of the resulting generation which contain both A and B, A but not B, B but not A, neither A nor B.

34. (1) In how many ways can 6 people be seated at a 'round' table? (The term 'round' must be taken to mean, as in the phrase 'round table conference', that there is no distinction between one seat and another.)

(2) If there are 6 seats at a 'round' table, in how many ways can 3 persons seat themselves?

(3) In how many ways can 6 different keys be arranged on a ring?

(4) In how many ways can 6 keys, of which 3 are alike and the rest different, be arranged on a ring?

(5) A molecule of benzene (C_6H_6) can be represented by a hexagon, and substitution products can be represented by placing appropriate letters or groups of letters at the vertices of the hexagon; thus a di-substitution product can be formed in 3 ways, according as the two vertices are 1, 2 or 3 sides of the hexagon apart. In how many ways can a tri-substitution product be formed, using 3 different atoms or groups?

35. A molecule of anthracene ($C_{14}H_{10}$) is represented by 3 hexagons placed in a line, each having one side in common with its neighbour or neighbours. Substitution products can be represented by placing appropriate letters at any of the 10 vertices not common to 2 hexagons. In how many ways can a di-substitution product be formed, using 2 identical atoms?

VI. THE BINOMIAL THEOREM
(for a positive integral index)

6.1. Pascal's triangle.

$$
\left.\begin{aligned}
(1+x)^0 &= 1 \\
(1+x)^1 &= 1+x \\
(1+x)^2 &= 1+2x+x^2 \\
(1+x)^3 &= 1+3x+3x^2+x^3 \\
(1+x)^4 &= 1+4x+6x^2+4x^3+x^4
\end{aligned}\right\} \tag{1}
$$

and so on.

The coefficients in these expansions form what is known as 'Pascal's Triangle'. If the triangle is arranged as follows:

```
              1
          1       1
      1       2       1
    1     3       3     1
  1     4     6     4     1
1     5    10    10     5     1
```

it will be seen that each number is the sum of the nearest two numbers of the row above. This is clearly due to the fact that each line of (1) is formed from the previous line by multiplying by $(1+x)$. The property provides an easy means of continuing the triangle, row by row, as far as we please. But it is also desirable that we should be able to write down the expansion in the general case, $(1+x)^n$.

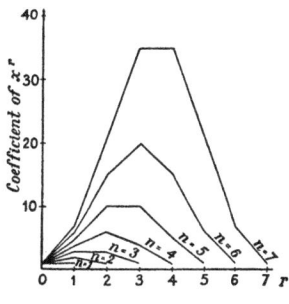

Fig. 6.1

6.2. The binomial theorem for a positive integral index.

Consider the product

$$(1+x)(1+x)(1+x)\ldots \text{to } n \text{ factors.}$$

To form a term x^r we must take the x from r of the n brackets, and the 1 from the remaining $n-r$ brackets. But we can choose r of the n brackets in $_nC_r$ ways: so there will be $_nC_r$ such terms.

$$\therefore \; (1+x)^n = {_nC_0} + {_nC_1}x + {_nC_2}x^2 + \ldots + {_nC_r}x^r + \ldots + {_nC_n}x^n, \tag{2}$$

$$= 1 + nx + \frac{n(n-1)}{2!}x^2 + \ldots + \frac{n(n-1)\ldots(n-r+1)}{r!}x^r + \ldots + x^n. \tag{3}$$

This result is known as the Binomial Theorem for a positive integral index. It should be noted that the expansion contains $n+1$ terms, and that so long as n is a positive integer, the general term

$$\frac{n(n-1)\ldots(n-r+1)}{r!}x^r$$

vanishes if $r > n$. If n is fractional or negative, the series does not come to an automatic end in this way at the $(n+1)$th term. The expansion in fact is different in form, being infinite instead of finite, the theorem is no longer true without limitation, and when true, is very much harder to prove. The conditions under which it is true, and some of the resulting expansions, will be considered in Chapter IX.

6.21. Special cases.

By giving n the values $0, 1, 2, 3, 4, \ldots$ we can see that the equations (1) may be reproduced. If a particular expansion, e.g. $(1+x)^{10}$, is required, the coefficients may be quickly obtained by writing down $\frac{10\ldots}{1\ldots}$, then adding the next factor in numerator and denominator, $\frac{5\!\!\!/10 \cdot 9\ldots}{1 \cdot 2\!\!\!/\ldots}$, and so on, cancelling down at each step.

6.22. Sum of the coefficients.

By putting $x = 1$ in the equation (2) we obtain the result

$$2^n = {_nC_0} + {_nC_1} + {_nC_2} + \ldots + {_nC_r} + \ldots + {_nC_n},$$

which was proved by another method in §5.6.

6.23. Expansion of $(1-x)^n$.

In the expansions (2) and (3), x may have any positive or negative value, but it is convenient to write down separately the expansion of $(1-x)^n$. By changing x into $-x$ in equation (3) we have:

$$(1-x)^n = 1 - nx + \frac{n(n-1)}{2!}x^2 - \ldots + (-1)^r \frac{n(n-1)\ldots(n-r+1)}{r!}x^r$$
$$+ \ldots + (-1)^n x^n. \quad (4)$$

6.3. Expansion of $(a+x)^n$.

It can similarly be shown that when n is a positive integer,

$$(a+x)^n = a^n + na^{n-1}x + \frac{n(n-1)}{2!}a^{n-2}x^2$$
$$+ \ldots + \frac{n(n-1)\ldots(n-r+1)}{r!}a^{n-r}x^r + \ldots + x^n. \quad (5)$$

6.4. Ratio of successive terms.

The general term given in the above expansions is the $(r+1)$th term in each case. Bearing in mind that each coefficient is formed from the previous one by adding one factor to the numerator and one to the denominator, we see that

$$\frac{\text{the } (r+1)\text{th term}}{\text{the } r\text{th term}} = \frac{n-r+1}{r}x.$$

If x is positive, this ratio will usually be greater than 1 for small values of r, and will remain greater than 1 as long as

$$(n-r+1)x > r,$$

i.e. as
$$nx + x > r + rx,$$

or as
$$(n+1)x > r(1+x),$$

or as
$$\frac{(n+1)x}{1+x} > r.$$

For example, if $n = 13$ and $x = \frac{2}{3}$, the ratio is greater than 1 as long as $\frac{14 \times \frac{2}{3}}{1+\frac{2}{3}} > r$, i.e. as $\frac{28}{5} > r$. In this case, therefore, $r = 5$ is the last value of r for which the ratio exceeds 1. This tells us that the 6th term is greater than the 5th, but that afterwards the terms diminish. Hence the 6th term (i.e. the term in x^5) is the greatest.

EXAMPLES VI

1. Write down Pascal's Triangle as far as the 10th row.

2. Obtain by the method of §6.21 the expansions of

(1) $(1+x)^8$.   (2) $(1+3x)^5$,   (3) $(1-x/2)^6$,   (4) $(2x-3y)^4$.

3. Find the sum of the coefficients in each of the expansions in Ex. 2.

4. What is the ratio (1) of the $(r+1)$th term to the rth, (2) of the $(r+1)$th coefficient to the rth, in the expansion of $(1+2x)^{12}$?

5. Find (1) which term is greatest, (2) which coefficient is greatest, in the expansion of $(1+x/3)^{16}$, when $x = 2$.

6. Find the greatest term in the expansion of
$$(1)\ (1+\tfrac{1}{5})^{20}, \qquad (2)\ (2+5)^{10}.$$

7. In the expansion of $(1+x)^{15}$, (1) state the number of terms, (2) show that the two middle coefficients are equal, and find them.

8. Show that when n is odd, the coefficients of the two middle terms in the expansion of $(1+x)^n$ are equal.

9. Write down the first 4 terms of the expansion of $(1+0\cdot02)^8$, and hence evaluate $(1\cdot02)^8$ to 3 places of decimals.

10. Evaluate $(0\cdot97)^6$ to 4 places of decimals.

11. Write down the first 4 terms of the expansion of $(1+x)^{20}$, state the total number of terms, the sum of the coefficients, and find the greatest term or terms when $x = \tfrac{1}{6}$.

12. By putting $x = 1$ and $x = -1$ in the expansion of $(1+x)^{20}$, show that the sum of the coefficients of the odd powers of x is equal to 2^{19}.

13. Show that if x is so small that terms in x^3 and higher powers can be neglected, $(1+x)(1+2x)^{10} = 1+21x+200x^2$ approximately. Find also the term containing x^3 and its approximate value when $x = 0\cdot01$.

14. Show that, if x is a small quantity, $(1+2x)(1-x)^6$ is approximately equal to $1-4x+3x^2$, and that, if $x = 0\cdot1$, the sum of the next two terms is less than $0\cdot01$.

15. If y denotes $x+1/x$, express x^7+1/x^7 in the form
$$y^7 + \mathsf{A}y^5 + \mathsf{B}y^3 + \mathsf{C}y,$$
where A, B, C are numerical coefficients.

16. If y denotes $x+1/x$, prove that $(x^7-1/x^7)\div(x-1/x)$ can be expressed in the form $y^6 + \mathsf{A}y^4 + \mathsf{B}y^2 + \mathsf{C}$, and find the values of A, B, C.

17. The amount of £1 at 3 per cent compound interest after n years is £$(1+0\cdot03)^n$. Use the Binomial Theorem to evaluate this formula for $n = 10$, correct to 2 places of decimals.

18. If 20 dice are tossed together, the chances of $0, 1, 2, 3, \ldots 20$ 'sixes' appearing are represented by the terms of the expansion of $(\tfrac{5}{6}+\tfrac{1}{6})^{20}$. Find the most probable number of 'sixes', and the chance of exactly that number of 'sixes' appearing.

VII. PROBABILITY

7.1. Introduction.

Probability may be regarded as a branch either of logic or of mathematics. The former aspect was considered by Leibniz in a dissertation (1669) on the mode of electing the Kings of Poland; by the authors of the *Port Royal Logic* (1662); by Locke and Hulme; and more recently, among others, by J. M. Keynes. But more attention has been given to the mathematical theory, which originated with the study of games of chance. Cardan, a gambler himself, wrote a small paper *De Ludo Aleae*, discussing the chance of throwing a particular number with two or three dice. He also considered the division of stakes between two players in an unfinished game, but his answer was wrong. Galileo showed that with three dice, there were more ways of throwing 10 than 9. The subject is more usually dated, however, from a correspondence between Pascal and Fermat (c. 1653) on the problem of the division of stakes, proposed to Pascal by the Chevalier de Méré, a well-known gambler. The real founder of mathematical probability was Jacob Bernoulli (1654–1705), who put forward what has been called the 'Principle of Indifference': that is to say, he regarded events as equally probable when he had no reason for preferring one to another. This was exactly opposite to the view of the logicians, who based probability on experience rather than on ignorance. The principle was developed by Laplace, whose 'Law of Succession' purported to measure numerically the influence of any experience, however limited. In the words of J. M. Keynes, 'No other formula in the alchemy of logic has exerted more astonishing powers. For it has established the existence of God from the premiss of total ignorance; and it has measured with numerical precision the probability that the sun will rise tomorrow.' Nevertheless the great authority of Laplace left the 'Principle of Indifference' in a strong position, and though it has since been criticised and qualified, it cannot be said to have been abandoned.

In the meantime the mathematical theory had been considerably developed, one of the contributors being de Moivre (1667–1754), who settled in England after the Revocation of the Edict of Nantes, taught mathematics in London, and was at one time chiefly dependent for his living on the solving of problems in probability in a tavern in St Martin's Lane. He wrote also on 'Annuities upon Lives', a subject of great interest at that time, when life-assurance had already been begun, though regarded by the public very much as a game of chance. Not until 1765 was the 'Equitable' founded, the first life-office to have a table of premiums calculated on a rational basis. (The method used had been proposed many years earlier by the astronomer Halley.) After 1800, actuarial theory developed more rapidly and became an important field of application for the ideas of probability.

The theory of probability has also been applied to errors of observation, sampling, and other statistical problems. More recently it has become important in physics. In the kinetic theory of gases, in thermodynamics and in quantum theory, there has emerged a new type of physical law, known as 'secondary' or 'statistical' in contrast to the 'primary' or 'deterministic' laws of classical physics. The movements of the molecules of a gas, and the flowing of heat from a hot body to a cold one are matters rather of probability than of necessity; and it has been said that the symbol ψ, used by quantum physicists, can be more nearly described as a probability than as anything else. In the words of Eddington, 'probability seems to have replaced aether as the nominative of the verb "to undulate"'.

7.2. Assumptions.

To attach a numerical measure to the probability of the occurrence of an event, we must first make some assumption as to certain ways in which it may or may not occur being 'equally likely'. Thus in the throwing of a die, it is usual to regard each of the six faces as equally likely to turn up, though a particular die may in fact be slightly biased in one direction or another. Or to take a less obvious case, suppose that a line is drawn at random from one corner A of a square ABCD, to cut the side BC at X. We might suppose that all points on BC are equally likely to be chosen, or we might suppose that all directions between AB and AC are equally likely to be chosen. These two assumptions are not the same, and would lead to different results.

The mathematical theory of probability seeks only to evaluate the probabilities (or chances) of certain events on the assumption that certain elementary events are equally likely. The definition of probability must therefore be in terms of such an assumption.

7.21. Definition.

If an event can happen in a ways and fail in b ways, all of which are equally likely, the 'probability' of its happening is defined as a/(a+b).

(Thus for a die, on the assumption mentioned above, the chance that on a given throw any particular face, say a six, will turn up is 1/6.)

The probability that the same event will fail to happen is, by the definition, $b/(a+b)$. (Thus with the die, the chance of not throwing a six is 5/6.) It will be observed that the sum of the probabilities that an event will happen and that it will fail to happen is 1.

7.3. Addition of probabilities.

If the probabilities of two mutually exclusive events (i.e. two events which cannot happen simultaneously) are p *and* q, *the chance that one or the other of them will happen is* p + q.

Thus the chance that when the die is thrown, either a 2 or a 6 will turn up is $\frac{1}{6} + \frac{1}{6}$, i.e. $\frac{1}{3}$, and the chance that 2, 4 or 6 will turn up is similarly $\frac{1}{2}$. (This latter result might also have been obtained from the assumption that odd and even throws are equally likely. The argument illustrates that the two assumptions (1) that all faces are equally likely, and (2) that odd and even throws are equally likely, are consistent with each other.)

7.31. Proof of the addition theorem.

If an event A can happen in a ways, and another event B in b ways, the two events being mutually exclusive (i.e. they cannot happen simultaneously), and if they can both fail in c ways, all these ways being equally likely, it is evident from the definition that

the chance of A happening is $a/(a+b+c)$,

the chance of B happening is $b/(a+b+c)$,

and the chance that either A or B will happen is

$$(a+b)/(a+b+c).$$

It should be noticed that the third result, like the first two, is obtained direct from the definition. It will also be observed that it is equal to the sum of the first two. We have in fact proved the theorem, provided only that the ways in which A and B can happen or fail are equally likely. The proof, however, is general, because the ways in which A and B can happen or fail can be subdivided into ways that are equally likely. (E.g. if A can happen in 3 ways out of 5 and B in 2 ways out of 7, it is equivalent to say that A can happen in 21 ways out of 35, B in 10 out of 35. $a+b+c$ is in fact the L.C.M. of the number of ways in which A can happen or fail and the number of ways in which B can happen or fail.)

7.32. Corollary.

We have already observed that the sum of the chances that an event will happen and that it will fail is 1. It therefore follows that a probability of 1 corresponds to certainty. (It should also be observed, that it is a direct consequence of the definition that probabilities must lie between 0 and 1.)

7.4. Multiplication of probabilities.

If two events whose probabilities are p *and* q *respectively are independent, then the probability that both should occur is* pq.

Thus if the die is thrown twice, the chance of a double six is $\frac{1}{36}$, and the chance that a six should not appear at either throw is $\frac{25}{36}$.

7.41. Proof of the multiplication theorem.

The multiplication theorem may be proved on the same lines as the addition theorem. Suppose that the first event can happen in a ways and fail in b ways, all equally likely, so that $p = a/(a+b)$, and that the second event can happen in a' ways and fail in b' ways, all equally likely, so that $q = a'/(a'+b')$. Then there are $(a+b)(a'+b')$ possible combinations, all equally likely, and of these there will be aa' combinations in which both events will happen. Hence the required probability is $aa'/(a+b)(a'+b')$, which is equal to pq.

7.5. It is instructive to see that if there are two sets of events, one set being independent of the other and one event of each set being bound to occur, then the total probability of all possible combinations is 1. If the probabilities of the first set are $p_1, p_2, p_3, \ldots$ and of the second set $q_1, q_2, q_3, \ldots$, where $p_1 + p_2 + p_3 + \ldots = 1$ and $q_1 + q_2 + q_3 + \ldots = 1$, the probabilities of the various possible combinations are, under the multiplication rule, the terms of the product $(p_1 + p_2 + p_3 + \ldots)(q_1 + q_2 + q_3 + \ldots)$, and their sum is therefore 1.

7.51. The method of § 7.5 is of very wide application. For example, if a pack of n cards is formed, x of them being clubs, y diamonds, z hearts and w spades, where $x + y + z + w = n$, and a card is drawn from the pack 'at random' (a phrase which implies that all cards are equally likely to be drawn), the chance of drawing a club is clearly $\frac{x}{n}$, of a heart $\frac{z}{n}$, of a diamond $\frac{y}{n}$ and of a spade $\frac{w}{n}$. If the card is replaced and the process is repeated twice (making three times in all), the probabilities of the various combinations are represented by the terms in the expansion of

$$\left(\frac{x}{n} + \frac{y}{n} + \frac{z}{n} + \frac{w}{n}\right)^3.$$

Thus the chance of drawing 3 club cards in succession is $\frac{x^3}{n^3}$; the chance of drawing 2 clubs and a diamond is $\frac{3x^2y}{n^3}$; and so on. If the pack is an ordinary one containing 13 cards of each suit, the probabilities are the terms of the expansion $(\frac{1}{4}+\frac{1}{4}+\frac{1}{4}+\frac{1}{4})^3$. In order to distinguish which of these fractions $\frac{1}{4}$ represent which suit, it is convenient to write the expression

$$\left(\frac{a}{4}+\frac{b}{4}+\frac{c}{4}+\frac{d}{4}\right)^3, \quad \text{where } a = b = c = d = 1.$$

The chance of drawing 2 clubs and a diamond is then $\dfrac{3a^2b}{64}$, which is equal to $\frac{3}{64}$.

7.6. Binomial probability-distribution.

In the case of two mutually exclusive events, one of which must occur at each of a number of trials, the expansion is a binomial one. For example, the chances of throwing 4, 3, 2, 1 or 0 sixes in 4 throws with a true die (the word 'true' implying that the six faces are all equally likely to appear at any throw) are given by the terms of the expansion $(\frac{1}{6}+\frac{5}{6})^4$.

For convenience this may be expanded as $\left(\dfrac{x+5}{6}\right)^4$ where $x = 1$, and the coefficients of x^4, x^3, etc. will represent the chances of 4 sixes, 3 sixes, etc. being thrown.

EXAMPLES VII

1. If a card is drawn at random from an ordinary pack, what is the chance that it will be either a knave, queen, king or ace?

2. What is the chance that a telephone number should be divisible (i) by 3, (ii) by 5, (iii) by 15, (iv) by either 3 or 5?

3. If 3 coins are tossed together, what is the chance that all 3 of them should be 'heads'? What is the chance that all 3 of them should be alike?

4. If two dice are thrown together, what is the chance that both should show even numbers? What is the chance that the total throw should be even?

5. If two dice are thrown together, what is the chance that the total throw should be (i) 2, (ii) 3, (iii) 4?

6. If 1 man in every 40 is colourblind, and 1 woman in every 40 is a carrier of the condition, what is the probability (i) that a colourblind man will marry a carrier woman, (ii) that a given marriage will involve both a colourblind man and a carrier woman?

7. A man has a case containing 10 cigarettes, of which 2 are Egyptian, 3 Turkish and the rest Virginian. If he takes one at random, what is the chance that it is an Egyptian? Supposing it is an Egyptian, what is the chance that the same thing will happen again when he takes another?

If he has 10 cigarettes, as stated at the beginning of the question and takes 3 in succession, what is the chance of their all being Turkish?

8. A man has 3 florins and 5 pennies in his pocket. If he takes out 3 of them at random, what is the chance of all 3 being pennies? If instead he takes out 3 in succession, replacing each before drawing the next, what is the chance of all 3 being pennies?

9. A man enters a dark room and wishes to take 2 penny stamps from a box containing 10 $1\frac{1}{2}d$. and 5 $1d$. ones all detached. If he takes 2 at random, what is his chance of getting what he wants? If he takes 3, what is the chance that at least 2 of them will be penny stamps?

10. Five cards are drawn from an ordinary pack, each being replaced before the next is drawn. Represent by a binomial expansion the chances that 5, 4, 3, 2, 1 or none of them should be hearts. Evaluate the chances of drawing (i) exactly 3 hearts, (ii) at least 3 hearts.

11. Four people play bridge together every day for a week, partners being determined each evening by cutting. What is the chance that Mr A should draw Mrs B as a partner on exactly 4 of the 7 occasions? What is the chance that it should happen on more than 4 of the 7 occasions?

12. A bridge player knows that his opponents hold 6 trumps between them. Assuming that each of these cards is equally likely to be in the hand of one opponent or the other, find the chance (i) that the 6 cards are equally divided, (ii) that one opponent, 'North', has 4 and the other, 'South', has 2, (iii) that the 6 cards are divided into 4 and 2.

(It may be remarked that the assumption proposed need not, and often should not, be made. There is often good reason for supposing that the cards are distributed in a particular way.)

13. A man travels daily by a train consisting of equal numbers of first, second and third class carriages. If on 5 successive days he chooses a carriage at random, write down an expression whose expansion would give the chances of his choosing various combinations of first, second and third class. What would be his chance of travelling first class on 4 of the 5 days?

14. In a game of Picquet there are 32 cards, of which 12 are dealt to each hand. Of the remaining 8 cards the holder of the 'elder hand' may then draw 5. Assuming that he does so, prove that the odds are 3 to 1 against his drawing any particular card not already in his hand. If the holder of the 'younger hand' then draws 3 cards, find the corresponding odds in his case.

15. If two cards are drawn from an ordinary pack of 52, find the chance that one should be an ace and the other a ten, knave, queen or king.

16. A bookmaker offers to take bets at the following odds for a race in which 6 horses are running:

Delphi 5 to 1 (corresponding to a $\frac{1}{6}$ chance that the horse will win)

Earnalot 3 to 1

Upjenkins 5 to 1

Royal Banner 4 to 1

Mount Vernon 1 to 1

Van Dieman 8 to 1.

Show that the sum of the corresponding chances is greater than 1, and explain why this is so. The bookmaker tries to arrange that the total of his possible gains and losses on each horse is the same. Show that if he succeeds in doing this, he will make the same profit whichever horse wins.

17. There are 3 horses in a race. A bookmaker offers to take 6 to 4 against anyone who will forecast the order in which the horses will finish. How many possible orders are there, and how much profit will the bookmaker secure if £12 is staked by backers on each of them? Will he make a profit if the backers all place the winner first but are equally divided as regards the order of the other two?

18. A botanist collects 15 plants of a certain species, and of these 15 plants 11 display a feature which the other 4 lack. He wishes to consider whether this is consistent with the hypothesis that $\frac{2}{3}$ of such plants display the feature and $\frac{1}{3}$ do not. If the chance of any plant displaying the feature were $\frac{2}{3}$, what would be the chance of obtaining 10 with it and 5 without it in a sample of 15? What would be the chance of obtaining 11 with the feature and 4 without it? Do you think that the observation of 11 with the feature is evidence against the hypothesis of a $\frac{2}{3}$ proportion?

19. There are six teams entered for a 'knock-out' competition. Two draw a bye in the first round, the byes being placed one in each half of the draw. Assuming that each team plays true to form, what chance has the second best team of reaching the final, and what chance has the third-best team of reaching the semi-final round?

20. An insurance policy provides for the payment of £100 at the end of one year if either of two persons, x aged 50, and y aged 60, die within the year. The chance that a life aged 50 should survive a year is 0·984, and the chance for a life aged 60 is 0·970. Find the probability that the £100 will be payable at the end of the year.

What would the probability have been if the £100 had been payable (*a*) only in the event of both persons dying during the year? (*b*) only in the event of *one* death during the year?

(Obtain these 3 answers independently of one another and check by considering the relation between them.)

21. Using the data of the above example, find the chance that of 3 lives aged 60 just one will survive for a year.

22. Out of every 1000 births, the following table shows the number of deaths occurring, on the average, in each year of age:

Age	No. of deaths	Age	No. of deaths
0–1	120	5–6	4
1–2	30	6–7	3
2–3	11	7–8	3
3–4	7	8–9	2
4–5	5	9–10	2

Find the chance at birth of surviving 10 years. Find also the chance at birth of dying in the 4th year of age, and the chance for a child aged exactly 3 of dying within a year.

23. Of 1000 entrants at age 20 to the staff of a big firm, the following table shows the numbers, on the average, who die or leave the service in each year of age:

Age	Deaths	Withdrawals
20–21	5	90
21–22	4	78
22–23	4	65
23–24	3	56
24–25	3	44
25–26	2	33

Make a table showing the number in the service at each age. Find (1) the probability that a particular entrant at age 20 will be still in the service at age 26, (2) the probability that a member of the staff aged 22 will withdraw between the ages of 24 and 26, and (3) the probability that a member aged 22 will no longer be in the service at age 25.

24. Of a given number of bachelors alive at age 30, the following table shows the number alive, on the average, at ages 30, 31, 32, and the numbers dying and marrying between 30 and 33.

Age	No. of living bachelors	Age	Decrements caused by	
			Bachelors dying	Bachelors marrying
30	536,664	30–31	4181	39,240
31	493,243	31–32	4323	36,138
32	452,782	32–33	4635	32,982

Find the probabilities:

(1) that a bachelor aged 30 will die unmarried before reaching age 31,

(2) that a bachelor aged 30 will marry within the next 3 years,

(3) that a bachelor aged 30 will fail to survive still unmarried to the age of 32,

(4) that a bachelor aged 30 will marry between the ages of 31 and 33.

25. If an area 1 mile square contains 10 barrage balloons, distributed at random, and a bomber whose wing span is 60 feet flies through at a level below that of the balloons, in a direction parallel to one of the sides of the square, what is the probability that it will hit one of the wires?

VIII. FINITE SERIES

8.1. Arithmetic and geometric progressions.

The series $a+(a+d)+(a+2d)+(a+3d)+...$ is called an 'Arithmetic Progression', the essential feature being that the difference between any term and the preceding term is constant.

The series $a+ar+ar^2+ar^3+...$ is called a 'Geometric Progression', the essential feature being that the ratio of any term to the preceding term is constant.

The two types may be contrasted by means of diagrams, as shown below, the heights of the columns representing the terms of the progressions.

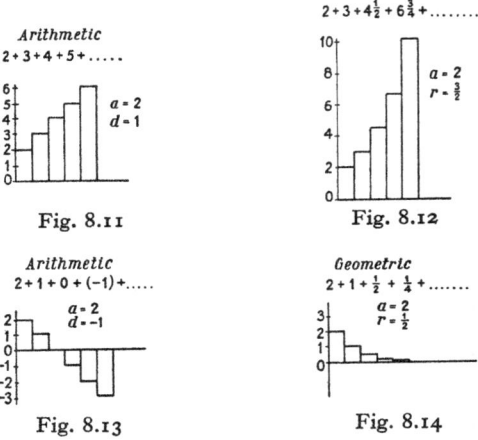

Fig. 8.11

Fig. 8.12

Fig. 8.13

Fig. 8.14

It will be noticed that the top left-hand corners (or top right-hand corners) of the columns representing terms of an Arithmetic Progression lie on a straight line, whereas those of a Geometric Progression lie on a curve of the type $y = ar^x$.

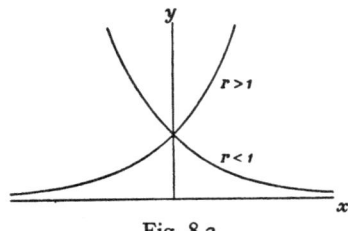

Fig. 8.2

8.2. Sum of an arithmetic progression.

Let $\qquad S = a+(a+d)+(a+2d)+\ldots+l,$ $\qquad\qquad$ (1)

where l is the last term. It will be seen that the third term is $a+2d$, the fourth $a+3d$, and so on; so if there are n terms altogether, the nth term, l, is given by

$$l = a+(n-1)d. \qquad\qquad (2)$$

Now $\qquad\qquad S = a+(a+d)+(a+2d)+\ldots+l.$

Reversing this, $\qquad S = l+(l-d)+(l-2d)+\ldots+a.$

Adding, $\qquad 2S = (a+l)+(a+l)+(a+l)+\ldots+(a+l)$
$\qquad\qquad\qquad = n(a+l).$

Therefore $\qquad\qquad S = \tfrac{1}{2}(a+l)n. \qquad\qquad (3)$

This result should be compared with the well-known formulae $A = \tfrac{1}{2}(a+b)h$, for the area of a trapezium, and $s = \tfrac{1}{2}(u+v)t$, for the distance covered by a body moving with uniform acceleration, both of which could in fact be derived from it by a process involving a limit.

Substituting for l from equation (2), we may also write

$$S = \tfrac{1}{2}n\{2a+(n-1)d\}. \qquad\qquad (4)$$

8.21. Special case.

The sum of the first n whole numbers is $\tfrac{1}{2}n(n+1)$.

8.3. Sum of a geometric progression.

Let $\qquad\qquad S = a+ar+ar^2+\ldots+ar^{n-1},$

where n is the number of terms.

Then $\qquad\qquad Sr = ar+ar^2+ar^3+\ldots+ar^n.$

Subtracting, $\qquad\qquad S-Sr = a-ar^n.$

Therefore $\qquad\qquad S = a(1-r^n)/(1-r).$

8.4. The method of differences.

The nature of an unknown series may often be discovered by tabulating the differences between successive terms, and, if necessary, repeating the process.

E.g. Consider the series $9 + 28 + 65 + 126 + 217 + 344 + ...$

First differences are	19	37	61	91	127
Second differences are		18	24	30	36
Third differences are			6	6	6

The reader should observe the numerous relations which can be read off from such a table, e.g.:

$$28 = 9 + 19,$$
$$65 = 9 + 19 + 37,$$
$$91 = 19 + 18 + 24 + 30,$$
$$19 + 37 + 61 + 91 + 127 = 344 - 9,$$
$$n\text{th first difference} = 19 + (18 + 24 + ... \text{ to } n - 1 \text{ terms})$$
$$= 19 + \tfrac{1}{2}(n-1)\{36 + (n-2)6\}$$
$$= 19 + \tfrac{1}{2}(n-1)(6n+24)$$
$$= 3n^2 + 9n + 7.$$

The nth term of the series $= 9 + \sum_{r=1}^{n-1} (3r^2 + 9r + 7).$

8.411. First differences and summation of some standard series.

We already know that for an Arithmetic Progression, the first difference is constant (d).

For a Geometric Progression, the first differences are also in G.P., with the same common ratio.

Series	a	ar	ar^2	...	ar^{n-1}	ar^n
Differences	$a(r-1)$	$ar(r-1)$	$ar^2(r-1)$	...	$ar^n(r-1)$	

This fact can be used as an alternative method for finding the sum of a G.P., for we see from the table that

$$a(r-1) + ar(r-1) + ... + ar^{n-1}(r-1) = ar^n - a.$$

Therefore $a + ar + ... + ar^{n-1} = a(r^n - 1)/(r-1).$

8.412. The series Σr, $\Sigma r(r+1)$, $\Sigma r(r+1)(r+2)$, **etc.**

These have as their first differences the terms of the series

$$\Sigma 1, \quad 2\Sigma r, \quad 3\Sigma r(r+1), \text{ etc.}$$

For example,

$$\sum_{r=0}^{n} r(r+1)(r+2) = 0.1.2+1.2.3+2.3.4+3.4.5$$
$$+ \ldots + n(n+1)(n+2).$$

First differences are

$$1.2(3-0), \quad 2.3(4-1), \quad 3.4(5-2), \quad \ldots, \quad (n+1)\{(n+2)-(n-1)\};$$

i.e. $3.1.2, \quad 3.2.3, \quad 3.3.4, \quad \ldots, \quad 3n(n+1);$

i.e. $3\Sigma r(r+1).$

From this it is obvious that

$$3\{1.2+2.3+3.4+\ldots+n(n+1)\} = n(n+1)(n+2)-0.1.2.$$

Therefore $1.2+2.3+3.4+\ldots+n(n+1) = \frac{1}{3}n(n+1)(n+2)$.

In the same way the sums of the series

$$\sum_{r=1}^{n} r, \quad \sum_{r=1}^{n} r(r+1), \quad \sum_{r=1}^{n} r(r+1)(r+2), \quad \text{etc.}$$

are $\frac{1}{2}n(n+1), \quad \frac{1}{3}n(n+1)(n+2), \quad \frac{1}{4}(n+1)(n+2)(n+3), \quad \text{etc.}$

The reader should note the close analogy between the above results and those obtained by differentiating and integrating the functions x, x^2, x^3, etc. Indeed the Calculus of Finite Differences, of which this work is a part, bears a close resemblance to the Infinitesimal Calculus, the finding of differences being analogous to differentiation and the summing of a series to integration. The Calculus of Finite Differences is used a great deal in actuarial work, especially for interpolation in tables of known values and for numerical integration.

8.42. Methods of application.

The sum of the squares of the first n integers, $\sum\limits_{r=1}^{n} r^2$, can be found from the above results.

$$\sum_{r=1}^{n} r^2 = \sum_{r=1}^{n} \{r(r+1)-r\}$$
$$= \tfrac{1}{3}n(n+1)(n+2) - \tfrac{1}{2}n(n+1)$$
$$= \tfrac{1}{6}n(n+1)(2n+1).$$

Similarly, $\sum\limits_{n=1}^{n} r^3 = \sum\limits_{r=1}^{n} \{r(r+1)(r+2) - 3r(r+1) + r\}$
$$= \tfrac{1}{3}n(n+1)(n+2)(n+3) - n(n+1)(n+2) + \tfrac{1}{2}n(n+1)$$
$$= \tfrac{1}{4}n^2(n+1)^2.$$

Or again, the expression at the end of § 8.4 can be evaluated:

$$9 + \sum_{r=1}^{n-1} (3r^2 + 9r + 7) = 9 + \sum_{r=1}^{n-1} \{3r(r+1) + 6r + 7\}$$
$$= 9 + (n-1)n(n+1) + 3(n-1)n + 7(n-1)$$
$$= n^3 + 3n^2 + 3n + 2.$$

The series in § 8.4 is therefore $\sum\limits_{r=1}^{n} (r^3 + 3r^2 + 3r + 2)$, and its sum could be found by another application of the same method.

8.5. Arithmetic and geometric means.

If a, x, b are in A.P., x is said to be 'the arithmetic mean of a and b'; if a, x, y, ..., b are in A.P., x, y, ... are said to be 'arithmetic means between a and b.'

If x is the arithmetic mean of a and b, we have

$$b - x = x - a.$$

Therefore $\qquad x = \tfrac{1}{2}(a+b).$

So the arithmetic mean of two numbers is the same as their average. Similarly if there are n numbers a, b, c, ..., their average, namely

$$(a+b+c+...) n,$$

is said to be their 'arithmetic mean'.

Similar definitions hold good for the term 'geometric mean'. If x is the geometric mean of a and b, $b/x = x/a$, and therefore $x = \sqrt{(ab)}$. Similarly, for n numbers, the geometric mean is $\sqrt[n]{(abc...)}$.

8.51. The arithmetic mean of a set of positive numbers is greater than or equal to the geometric mean.

(i) For 2 numbers a, b:

We have to prove that

$$\tfrac{1}{2}(a+b) \geqslant \sqrt{(ab)}. \tag{1}$$

It is sufficient to prove that

$$\tfrac{1}{4}(a^2 + 2ab + b^2) \geqslant ab, \tag{2}$$

or that

$$a^2 + 2ab + b^2 \geqslant 4ab, \tag{3}$$

or that

$$a^2 - 2ab + b^2 \geqslant 0, \tag{4}$$

which is so, since the L.H.S. is a perfect square.

Equality can only occur if $a = b$.

(*Note.* The phrase 'it is sufficient to prove that' implies that the inequality (1) can be deduced from (2), (2) from (3), and (3) from (4). This is not true unless a and b are restricted to positive values. Consider, for example, the result of substitution -4 for a, -2 for b.)

(ii) For n numbers $a, b, c, \ldots$:

If the n numbers are all equal, the arithmetic mean is equal to the geometric mean. If not, let p and q be the greatest and least. If we replace each of these two numbers by $\tfrac{1}{2}(p+q)$, the sum of the n numbers will be unchanged, but the product will be increased, since $\{\tfrac{1}{2}(p+q)\}^2 > pq$. Therefore the A.M. (arithmetic mean) will be unchanged and the G.M. increased. This process can be repeated, and as it is continued, the numbers will approach equality. But if they were all equal, their A.M. and G.M. would be equal. Originally, therefore, the A.M. was greater than the G.M.

8.52. Corollaries.

If the sum of n positive numbers is constant, their product is greatest when they are all equal.

If the product of n positive numbers is constant, their sum is least when they are all equal.

EXAMPLES VIII

1. State the nature of each of the following series and find in each case the nth term and the sum to n terms:

$$(1) \quad 5+8+11+14+...,$$
$$(2) \quad 8+12+18+27+...,$$
$$(3) \quad 192+144+108+81+...,$$
$$(4) \quad 4+1-2-5-...,$$
$$(5) \quad 1+4+9+16+...,$$
$$(6) \quad 3+9+19+33+...,$$
$$(7) \quad 2+6+12+20+....$$

2. An A.P. of 40 terms begins with 18 and ends with 44. Find its sum and its second term.

3. A G.P. of 7 terms begins with 12 and ends with 768. Find its common ratio and its sum.

4. The first two terms of a series are a and b. Find the nth term and the sum to n terms, (1) if it is an A.P., (2) if it is a G.P.

5. Use the method of §8.3 to find the sum of 20 terms of the series
$$1+2x+3x^2+4x^3+....$$

6. Find the sum of n terms of the series
$$a+(a+d)x+(a+2d)x^2+(a+3d)x^3+....$$

7. Evaluate:

$$(1) \quad \sum_{r=1}^{n}(a+rd), \qquad (2) \quad \sum_{r=0}^{n}ab^{-r},$$
$$(3) \quad \sum_{r=1}^{n}(r^2+3r), \qquad (4) \quad \sum_{r=1}^{n}(r^3+3r^2+2).$$

8. Find the nth term and the sum to n terms of the series
$$1+9+29+67+129+221+....$$

9. Find the first differences of the series
$$\frac{1}{1.2}+\frac{1}{2.3}+\frac{1}{3.4}+...+\frac{1}{n(n+1)}.$$

Hence, evaluate
$$\frac{1}{1.2.3}+\frac{1}{2.3.4}+...+\frac{1}{n(n+1)(n+2)}.$$

10. Find the differences between the $(r+1)$th and the rth terms for each of the series $\Sigma 1/r$, $\Sigma 1/r(r+1)$, $\Sigma 1/r(r+1)(r+2)$. Hence, find the sum to n terms of $\Sigma 1/r(r+1), \Sigma 1/r(r+1)(r+2), \Sigma 1/r(r+1)(r+2)(r+3)$.

11. Find the sum of 20 terms of the series $1 + 1/1 \cdot 05 + 1/(1 \cdot 05)^2 + \ldots$.

12. After how many terms will each term of the series
$$1 + \tfrac{2}{3} + (\tfrac{2}{3})^2 + (\tfrac{2}{3})^3 + \ldots$$
be less than $0 \cdot 01$?

13. How many terms of the series $10 + 12 + 14 + \ldots$ must be taken to give a sum greater than 200?

14. How many terms of the series $1 \cdot 03 + (1 \cdot 03)^2 + (1 \cdot 03)^3 + \ldots$ must be taken to give a sum greater than 100?

15. If the sum of the first n terms of a series is given by the formula $an^2 + bn$, find the first two terms and the nature of the series.

16. Find the sum of all the odd numbers between 0 and 200, with the exception of those that are divisible by 5.

17. The formula for the sum of n terms of a certain geometrical progression is known to be $3^n - 1$ for all values of n. Find the first 3 terms.

18. If x, y, z are consecutive terms of a geometric progression, prove that $\log x$, $\log y$, $\log z$ are consecutive terms of an arithmetic progression.

19. The 8th term of an arithmetic progression is twice the 3rd term, and the sum of the first 8 terms is 39. Find the first 3 terms, and show that the sum to n terms is $3n(n+5)/8$.

20. A man received in salary (paid at the end of each year) £200 for the first year, £210 for the second year, £220 for the third, and so on up to a maximum of £350. His salary then continued unchanged at £350 per annum. How much in all had he received at the end of 25 years? Starting with the same salary of £200 per annum, what annual increment during the whole of the 25 years would have given him the same total sum at the end of the 25 years?

21. A triangle has a horizontal base of 8 cm., and height 7 cm. Lines are drawn parallel to the base at vertical intervals of $0 \cdot 05$ cm. Find their total length. (It is best to include the base itself, 8 cm., and the vertex, 0 cm., and subtract them at the end.)

22. A wooden pyramid on a square base of side 6 in. has a height of 12 in. It is sawn into slices 1 in. thick by a series of cuts parallel to the base. Find the total area of the cuts.

23. A golf ball is dropped on to a stone floor. It bounces and strikes the floor again after 2 sec. Each succeeding rebound occupies 0·6 of the time taken by the preceding one. How many times will the ball have hit the ground before the rebounds become shorter than $\frac{1}{20}$ sec.? Find the total time between the first and last of these hits.

24. Observe that in the attached figure an equilateral triangle with its vertex above its base can be formed by taking as ends of its base any two points on the same level. How many such triangles will there be? If a similar figure were drawn with a base having n sections instead of 7, how many such triangles would there be?

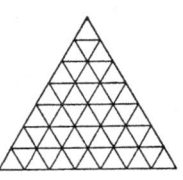

Fig. 8.3.

25. A set of wooden balls have radii 1, 2, 3, ..., 10 cm. Find their total volume.

26. If compound interest is reckoned at r per cent per annum, a payment of £P made now is worth £P$(1+r/100)^n$ in n years time. Find the value of a series of 10 annual payments of £50 at the time of the last payment, reckoning compound interest at 3 per cent per annum. ($\log 1\cdot03 = 0\cdot012837$.)

27. If compound interest is reckoned at r per cent per annum, the present value of a payment of £A due n years hence is £A$/(1+r/100)^n$. Find the present value of a series of 12 annual instalments of £100, the first one being due 1 year hence, reckoning compound interest at 5 per cent per annum. ($\log 1\cdot05 = 0\cdot021189$.)

28. A debt of £100 is to be cleared by 5 annual instalments, the first being 1 year after the debt is contracted. Compound interest is reckoned at 4 per cent per annum. Find how large the instalments should be. ($\log 1\cdot04 = 0\cdot017033$.)

29. Find the present value of an annuity of £a per annum for n years, the first annual payment being immediate, reckoning compound interest at r per cent per annum.

30. Prove that the largest rectangular box having a given surface area is a cube.

31. If two bodies of unit mass have momenta v_1 and v_2, and the sum of their momenta is constant, prove that their kinetic energy $\frac{1}{2}v_1^2 + \frac{1}{2}v_2^2$ is least when $v_1 = v_2$. (*Hint.* The kinetic energy = a constant $- v_1 v_2$, and is therefore least when $v_1 v_2$ is greatest.)

IX. INFINITE SERIES

9.1. Introduction.

The idea of an infinite series is a natural development from decimal fractions, which were invented in the 16th century. The digits of a decimal are a series of terms by which we obtain closer and closer approximations to a desired number. Some decimals, such as ·11111... and 3·14159... may be continued indefinitely, the successive approximations becoming closer and closer to the desired number, but never reaching it exactly. Thus decimals lead to the idea of infinite series.

Logarithms were invented by Napier in 1614, but were not at first expressed as decimals nor calculated by the aid of infinite series.

Soon after this, however, about the time of the invention of the calculus, series came into prominence. Gregory St Vincent (1584–1667) obtained an infinite geometric progression and applied the idea to Zeno's *Achilles* paradox; in 1668, Nicholas Mercator and James Gregory obtained substantially the two series for $\log(1+x)$ and $\frac{1}{2}\log\{(1+x)/(1-x)\}$ respectively, both these series being stated in general terms by Wallis in 1695; Newton in 1676 stated the Binomial Theorem for any index, and he also used series for the solution of differential equations. James Gregory obtained the series for $\tan^{-1}x$ and $\pi/4$. In 1712 Taylor gave his general theorem for the expansion of functions in the form of series, though without any proof of value.

By modern standards, none of these results had been proved. Nor was it fully realised that to be of use a series must give successive approximations converging towards a desired value. Euler (1707–1783) occasionally warns his readers against the use of series now called 'divergent' which do not satisfy this requirement. Nevertheless Euler himself was by no means careful in this respect, writing, for example, '$1-3+5-7+... = 0$', and '$...+1/n^2+1/n+1+n+n^2+... = 0$' (obtained by summing two infinite geometric progressions and adding the results). Leibniz put x equal to 1 in the expansion $1/(1+x) = 1-x+x^2-...$, obtaining

$$\tfrac{1}{2} = 1-1+1-1+...,$$

and Guido Grandi concluded that $\frac{1}{2} = 0+0+0+...$, a result that appeared to him to illustrate the creation of the world out of nothing.

Speaking generally, series were much used in the 18th century, but without sufficient regard to the question of convergence. A few writers, notably Waring, d'Alembert and Nicolaus Bernoulli (the one born in 1695), held stricter views. Lagrange, in his earlier years, used series without due care, but his later works were written at the beginning of the 19th century, a period when new standards of criticism were beginning to appear. Gauss and Cauchy gave strictly rigorous treatments of series, and Abel, a Norwegian mathematician, was an outspoken critic of the older methods. It is recorded that after hearing a lecture by Cauchy, Laplace hurried home and remained in seclusion until he had verified the convergence of all the series used in his great work *Mécanique Céleste*. Fortunately they were all convergent.

The usefulness of an infinite series lies in the fact that it expresses a function by means of a succession of closer and closer approximations. This is convenient not only for obtaining numerical values but for finding algebraical approximations and for performing operations on otherwise intractable functions. But clearly the essential condition is that the successive approximations given by taking more and more terms of the series should in fact converge towards the desired value. Hence arises the necessity for tests of convergence.

9.2. Infinite geometric series.

Consider the series $1 + \frac{1}{2} + \frac{1}{4} + \frac{1}{8} + \dots$. Adding the terms together, we obtain in succession $1, 1\frac{1}{2}, 1\frac{3}{4}, 1\frac{7}{8}, \dots$. The more terms we take, the nearer does the sum approach 2. By taking a sufficient number of terms we can make the sum as near 2 as we please. Such a series is said to be 'convergent', and this particular series is said to 'converge to the sum 2'. (We may also say, 'The sum "approaches" 2', and this may be written 'The sum $\to 2$'.)

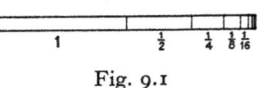

Fig. 9.1

Similar remarks apply to the recurring decimal $0 \cdot 9999 \dots$. By taking a sufficient number of places we can make the value as near 1 as we please. The statement '$0 \cdot \dot{9} = 1$' must be taken as a short way of saying 'The series $0 \cdot 9 + 0 \cdot 09 + 0 \cdot 009 + \dots$ converges to the sum 1'.

These are both examples of geometric progressions whose common ratios are less than 1. If we take a progression whose common ratio is greater than 1, e.g. $1 + 2 + 4 + 8 + \dots$, we find there is not any particular number which the sum approaches as we take more and more terms. In fact by taking a sufficient number of terms we can make the sum as large as we please. Such a series is said to be 'divergent'. (We may also say, 'The sum "approaches infinity"', and this may be written, 'The sum $\to \infty$'.)

9.21. 'Sum to infinity' of a geometric series.

It was proved in Chapter VIII that the sum s_n of n terms of the geometric progression $a + ar + ar^2 + \dots$ is given by $a(1 - r^n)/(1 - r)$. If r is 'numerically less than 1', i.e. if $-1 < r < 1$, $r^n \to 0$ as $n \to \infty$. (I.e. by making n large enough we can make r^n as near 0 as we please.) The sum then approaches $a(1-0)/(1-r)$, i.e. $a/(1-r)$.

The series is said to 'converge to the sum $a/(1-r)$'. $a/(1-r)$ is sometimes called the 'limiting sum' or the 'sum to infinity' of the series.

9.22. Divergent geometric series.

If r is greater than 1, $r^n \to \infty$ as $n \to \infty$. (I.e. we can make r^n as large as we please by taking a sufficiently large value of n.) The sum then $\to \infty$, and the series is divergent.

If $r < -1$, r^n oscillates between $+\infty$ and $-\infty$, and the series is again said to be 'divergent'.

9.3. Definition of convergence.

Let s_r denote the sum to r terms of an infinite series $u_1 + u_2 + u_3 + \dots$. If s_r approaches a limit s as $r \to \infty$, the series is said to 'converge to the sum s'.

9.31. It is easy to see that this cannot happen unless $u_r \to 0$ as $r \to \infty$. This condition is not, however, sufficient by itself to prove convergence, as will be seen from the following example:

$$1 + \tfrac{1}{2} + \tfrac{1}{3} + \tfrac{1}{4} + \tfrac{1}{5} + \tfrac{1}{6} + \tfrac{1}{7} + \tfrac{1}{8} + \dots$$
$$> 1 + \tfrac{1}{2} + (\tfrac{1}{4} + \tfrac{1}{4}) + (\tfrac{1}{8} + \tfrac{1}{8} + \tfrac{1}{8} + \tfrac{1}{8}) + \dots$$
$$= 1 + \tfrac{1}{2} + \tfrac{1}{2} + \tfrac{1}{2} + \dots.$$

The series $1 + \tfrac{1}{2} + \tfrac{1}{3} + \tfrac{1}{4} + \tfrac{1}{5} + \dots$ is therefore divergent.

9.32. D'Alembert's test for convergence.

One of the most useful tests for convergence is to compare the given series with a geometric series. For example, consider the series

$$1 + 2x + 3x^2 + 4x^3 + \dots.$$

The $(r+1)$th term is $(r+1)x^r$ and the ratio u_{r+1}/u_r is $\dfrac{r+1}{r}x$. As $r \to \infty$, this $\to x$.

First, suppose x is positive and less than 1. After a certain point, the ratio will be near enough to x for us to find a number k, between x and 1, such that u_{r+1}/u_r is always less than k from that point onwards (see figure). So the terms of the series,

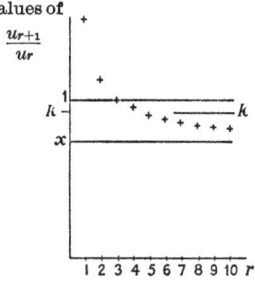

Fig. 9.2.

after that point, are less than those of a convergent geometric series of common ratio k. The given series is therefore convergent if $0 < x < 1$.

If $x > 1$, the terms, after a certain point, are greater than those of a divergent geometric series, and the given series is therefore divergent.

If $x = 1$, this test does not help us to prove either convergence or divergence.

If x is negative, we must again divide into two cases:

(i) If $-1 < x < 0$, the series can be expressed as the difference of two series, one containing the positive, the other the negative terms. Each of these is convergent, being part of the given series for the corresponding positive value of x, and so, therefore, will be the series with which we are dealing.

(ii) If $x < -1$, the terms will alternate in sign, but after a certain point will increase in numerical value. The series is therefore divergent, since u_n does not approach zero.

Summarizing, we may state d'Alembert's test as follows:

If u_{r+1}/u_r approaches a limit l as $r \to \infty$, then the series is convergent if l is numerically less than 1 (i.e. if $-1 < l < 1$), and divergent if l is numerically greater than 1. If $l = 1$, or -1, the test fails.

9.33. Series of alternately positive and negative terms.

We have seen that $1 + \frac{1}{2} + \frac{1}{3} + \frac{1}{4} + \frac{1}{5} + \dots$ is divergent. On the other hand $1 - \frac{1}{2} + \frac{1}{3} - \frac{1}{4} + \frac{1}{5} - \dots$ is convergent, as will be seen from the following diagram representing values of s_r.

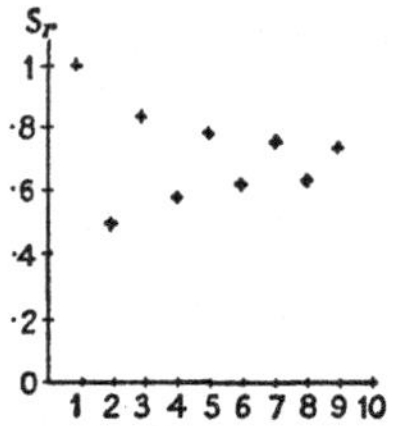

It will be observed that s_1, s_3, s_5, ... are a decreasing sequence of numbers, while s_2, s_4, s_6, ... are an increasing sequence, and that the two sequences approach each other. On these lines it is possible to prove the following test:

If the terms of a series are alternately positive and negative, and if $u_n \to 0$ as $n \to \infty$, the series is convergent.

Fig. 9.3

This test is sometimes useful for cases in which d'Alembert's test fails.

9.4. Binomial series.

When n is a positive integer, the series

$$1 + nx + \frac{n(n-1)}{1.2}x^2 + \dots + \frac{n(n-1)(n-2)\dots(n-r+1)}{r!}x^r + \dots$$

comes automatically to an end at the $(n+1)$th term, all succeeding terms being zero. It was proved in Chapter VI that the sum of the finite series is $(1+x)^n$. If, however, n is fractional or negative, the series is an infinite one and the method of proof used in Chapter VI is not applicable. But it is possible to prove, by more advanced methods, that when the series is convergent its sum is $(1+x)^n$.

It was seen in §6.4 that $\dfrac{u_{r+1}}{u_r} = \dfrac{n-r+1}{r}\,x$. As r increases, this approaches $-x$. Hence, by d'Alembert's test, the series is convergent if x is numerically less than 1 (i.e. if $-1 < x < 1$).

We shall therefore assume that

$$(1+x)^n = 1 + nx + \frac{n(n-1)}{1.2}x^2 + \ldots + \frac{n(n-1)\ldots(n-r+1)}{r!}x^r + \ldots$$

for all values of n, provided $-1 < x < 1$.

9.41. Special cases.

In special cases it is advisable first to substitute for n without simplifying, and to simplify afterwards as a separate step. Thus if $n = \frac{1}{2}$,

$$(1+x)^{\frac{1}{2}} = 1 + (\tfrac{1}{2})\,x + \frac{(\tfrac{1}{2})(-\tfrac{1}{2})}{1.2}x^2 + \frac{(\tfrac{1}{2})(-\tfrac{1}{2})(-\tfrac{3}{2})}{1.2.3}x^3$$
$$+ \frac{(\tfrac{1}{2})(-\tfrac{1}{2})(-\tfrac{3}{2})(-\tfrac{5}{2})}{1.2.3.4}x^4 + \ldots$$
$$= 1 + \tfrac{1}{2}x - \tfrac{1}{8}x^2 + \tfrac{1}{16}x^3 - \tfrac{5}{128}x^4 + \ldots,$$

provided $-1 < x < 1$.

The following special cases are worth remembering:

$$(1+x)^{-1} = 1 - x + x^2 - x^3 + \ldots,$$
$$(1-x)^{-1} = 1 + x + x^2 + x^3 + \ldots,$$
$$(1+x)^{-2} = 1 - 2x + 3x^2 - 4x^3 + \ldots,$$
$$(1-x)^{-2} = 1 + 2x + 3x^2 + 4x^3 + \ldots.$$

9.42. Expansion of $(a+x)^n$.

This must first be expressed as $a^n(1+x/a)^n$. It is then seen that the expansion

$$(a+x)^n = a^n + na^{n-1}x + \frac{n(n-1)}{2!}a^{n-2}x^2$$
$$+ \ldots + \frac{n(n-1)\ldots(n-r+1)}{r!}a^{n-r}x^r + \ldots$$

will be valid for all values of n, provided that $-a < x < a$.

9.43. Approximations.

From the expansion of $(1+x)^{\frac{1}{2}}$ we observe that when x is small, $\sqrt{(1+x)} \simeq 1 + \frac{1}{2}x$. This is a useful approximation and should be remembered.

More generally, $\sqrt[n]{(1+x)} \simeq 1 + x/n$, and similarly, $\sqrt[n]{(1-x)} \simeq 1 - x/n$.

9.44. Error in approximations.

If x is small, the first term neglected will usually give a good idea of the amount of the error. Or we may proceed as in the following examples:

From the equation given in §9.41 it is seen that the error in the approximation just given for $\sqrt{(1+x)}$ is $-\frac{1}{8}x^2 + \frac{1}{16}x^3 - \frac{5}{128}x^4 + \ldots$.

Let us suppose that x is positive. The error is then numerically less than $\frac{1}{8}x^2 + \frac{1}{16}x^3 + \frac{5}{128}x^4 + \ldots$. We know that the ratio $\dfrac{u_{r+1}}{u_r}$ is $\dfrac{n-r+1}{r}x$, which in this case is equal to $\dfrac{r-1\frac{1}{2}}{r}(-x)$. Its numerical value approaches x and is less than x. Hence the error is numerically less than $\frac{1}{8}x^2 + \frac{1}{8}x^3 + \frac{1}{8}x^4 + \ldots$, or $\frac{1}{8}x^2(1 + x + x^2 + \ldots)$, i.e. $\frac{1}{8}x^2/(1-x)$. As a numerical example, consider $\sqrt{(103)}$, which $= 10\sqrt{(1+0.03)}$, $\simeq 10(1 + 0.015)$, $= 10.15$, with an error whose numerical value is less than $\dfrac{10}{8}\cdot\dfrac{(0.03)^2}{0.97}$, which is less than 0.0012.

(Actually, the error is -0.0011, approximately.)

Or again, if we take $1 + 2x + 3x^2$ as an approximation for $(1-x)^{-2}$, the error is $4x^3 + 5x^4 + 6x^5 + \ldots$. Here the ratio u_{r+1}/u_r continually diminishes towards the value x. Hence the error is less than the sum of the geometric series beginning $4x^3 + 5x^4 + \ldots$, i.e. than

$$\frac{4x^3}{1 - 5x/4}, \quad \text{or} \quad \frac{16x^3}{4 - 5x}.$$

9.5. The exponential series.

The series $1 + x/1! + x^2/2! + x^3/3! + \ldots + x^r/r! + \ldots$ has remarkable properties. In the first place it is convergent for all values of x, for the $(r+1)$th term, namely $x^r/r!$, is obtained from the rth by multiplying by x/r, so the ratio $u_{r+1}/u_r = x/r$, which $\to 0$ as $r \to \infty$, whatever the value of x.

Further, it can be proved that its sum is e^x, where e is the value obtained by putting $x = 1$, namely

$$1 + 1 + 1/2! + 1/3! + \ldots, \quad \text{or} \quad 2.718\ldots$$

So we write
$$e^x = 1 + x + x^2/2! + x^3/3! + \ldots,$$

for all values of x. This result is known as the 'Exponential Theorem'.

Students of the calculus will know that $\dfrac{d}{dx} e^x = e^x$, and will find it interesting to see the result of differentiating this series term by term (a process not always justifiable, though it gives the correct result in this case).

9.51. Greatest term, etc.

The value, in the exponential series, of the ratio u_{r+1}/u_r is x/r, and if x is numerically greater than 1, this will at first be greater than 1; but as r increases it will become less than 1. Hence the terms will at first increase in numerical value and then decrease. If x is numerically less than 1, however, they will decrease from the first term onwards.

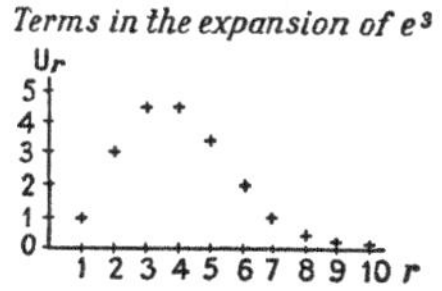

Terms in the expansion of e^3

Fig. 9.4

In the expansion of e^3, $u_{r+1}/u_r = 3/r$. This is greater than 1 if $r = 1$ or 2, equal to 1 if $r = 3$ (hence the 3rd and 4th terms are equal), and after that, less than 1. If the first 8 terms are taken as an approximation for e^3, the error, namely

$$3^8/8! + 3^9/9! + 3^{10}/10! + \ldots,$$

is less than the sum of the geometric series beginning

$$3^8/8! + 3^9/9! + \ldots,$$

i.e. less than $(3^8/8!)/(1 - \tfrac{3}{9})$, i.e. 0·244.

9.6. Logarithmic series.

It can be proved that

$$\log_e(1+x) = x - x^2/2 + x^3/3 - x^4/4 + \ldots + (-1)^{r-1} x^r/r + \ldots,$$

within the range of convergence. Using d'Alembert's test,

$$u_{r+1}/u_r = -rx/(r+1),$$

and this $\to -x$ as $r \to \infty$. Hence the series is convergent if $-1 < x < 1$. It is also convergent if $x = 1$, since the terms are alternately $+$ and $-$, and $u_r \to 0$ (§9.33). Hence

$$\log_e(1+x) = x - x^2/2 + x^3/3 - x^4/4 + \ldots + (-1)^{r-1} x^r/r + \ldots$$

if

$$-1 < x \leqslant 1.$$

Changing x into $-x$,

$$\log_e(1-x) = -x - x^2/2 - x^3/3 - x^4/4 - \ldots - x^r/r - \ldots$$

if

$$-1 \leqslant x < 1.$$

LA

5

These series converge too slowly to be of practical use for calculating logarithms, but a series converging much more rapidly can be obtained by subtracting one from the other:

$$\log_e\{(1+x)/(1-x)\} = 2\,(x+x^3/3+x^5/5+\ldots)$$

if $$-1<x<1.$$

This again may be usefully transformed as follows:

Let $$(1+x)/(1-x) = m/n.$$

Then $$n+nx = m-mx,$$

and $$x = (m-n)/(m+n).$$

Therefore

$$\log_e\frac{m}{n} = 2\left\{\frac{m-n}{m+n}+\frac{1}{3}\left(\frac{m-n}{m+n}\right)^3+\frac{1}{5}\left(\frac{m-n}{m+n}\right)^5+\ldots\right\},$$

and since $x = (m-n)/(m+n)$, x will be fractional (i.e. numerically less than 1) for all positive values of m and n. Hence the expansion will be valid for all positive values of m and n. It will be seen, however, that the series does not converge very rapidly unless m and n are nearly equal.

9.7. Other series.

By Maclaurin's Theorem in the differential calculus, or otherwise, many other expansions can be obtained. The following are among the most important:

$$\sin x = x - x^3/3! + x^5/5! - x^7/7! + \ldots \text{ for all values of } x,$$
$$\cos x = 1 - x^2/2! + x^4/4! - x^6/6! + \ldots \text{ for all values of } x,$$
$$\sinh x = x + x^3/3! + x^5/5! + x^7/7! + \ldots \text{ for all values of } x,$$
$$\cosh x = 1 + x^2/2! + x^4/4! + x^6/6! + \ldots \text{ for all values of } x,$$
$$\tan^{-1} x = x - x^3/3 + x^5/5 - x^7/7 + \ldots \text{ for } -1<x\leqslant 1.$$

It is important to notice that in the first two of these expansions the angle x is measured in radians. The series (known as 'Gregory's series') for $\tan^{-1} x$ will likewise give the angle in radians. For example, if we substitute 1 for x, we obtain $\tan^{-1} 1$, namely $\pi/4$. Therefore

$$\pi/4 = 1 - \tfrac{1}{3} + \tfrac{1}{5} - \tfrac{1}{7} + \ldots.$$

This series could be used for calculating the value of π, but it converges very slowly. It is more convenient to use smaller values of x. By means, for example, of 'Machin's formula',

$$\pi/4 = 4\tan^{-1}\tfrac{1}{5} - \tan^{-1}\tfrac{1}{239},$$

we can obtain π as the difference of two rapidly convergent series.

It should also be remarked that the above series can be used to define the sine, cosine, etc. as functions of a variable number x not necessarily connected in any way with an angle. These functions are important as modes of variation. In fact it has been shown that any periodic vibration can be expressed in terms of sines or cosines. In this way they have a significance beyond their trigonometrical meaning.

EXAMPLES IX

1. Find the sum of the following infinite series:

(1) $48 + 12 + 3 + \frac{3}{4} + \ldots$,

(2) $0\cdot\dot{4}$,

(3) $0\cdot\dot{2}\dot{1}$,

(4) $x/(1+x) + x^2/(1+x) + x^3/(1+x) + \ldots$,

(5) $axy + a^2x^2y + a^3x^3y + \ldots$,

stating, where necessary, the conditions of convergence.

2. If x and y are positive, $x+y = 1$, and

$$a = 1 + x + x^2 + x^3 + \ldots,$$
$$b = 1 + y + y^2 + y^3 + \ldots,$$
$$c = 1 + xy + x^2y^2 + x^3y^3 + \ldots,$$

prove that $\qquad ab = a+b,$

and $\qquad abc = a+b+c.$

3. Discuss the convergence of the following series:

(1) $n + (n+1)x + (n+2)x^2 + (n+3)x^3 + \ldots$,

(2) $1/(1+a) + 1/(1+a^2) + 1/(1+a^3) + \ldots$,

(3) $1!/x + 2!/x^2 + 3!/x^3 + \ldots$,

(4) $1/x - 1/(2x^2) + 1/(3x^3) - 1/(4x^4) + \ldots$.

4. Use the Binomial Theorem to expand the following functions as far as the 4th term:

(1) $(1+x)^{-3}$,　　　(2) $\sqrt{(1+2x)}$,　　　(3) $1/\sqrt[3]{(1+x^2)}$,

(4) $(2+x)^{-2}$,　　　(5) $\sqrt{(3-2x)}$,　　　(6) $1/(5-x)$.

State in each case the range of values of x for which the expansion is valid.

5. Expand $1/\{1-(x+x^2)\}$ as far as the term containing x^4. Also $1/(1+x+x^2)$.

6. Use the first two terms of an expansion to obtain an approximation for $\sqrt[3]{(1\cdot04)}$ and show that the third term would not affect the third place of decimals.

7. Find to 4 places of decimals approximations for:

(1) $\sqrt{(1\cdot03)}$,   (2) $(1\cdot03)^{-\frac{1}{3}}$,   (3) $\sqrt[3]{65}$,

(4) $(4\cdot08)^{\frac{3}{2}}$,   (5) $(0\cdot96)^{-3}$,   (6) $\sqrt{98}$.

8. Prove that when x is small, $1/(1-x)$ is approximately equal to $1+x+x^2$, with an error equal to $x^3/(1-x)$.

9. Find the ratio of the seventh term to the sixth term in the expansion of $(1+0\cdot05)^{13}$ by the binomial theorem. Prove that the sum of all the terms after the sixth is less than $0\cdot00003$.

10. Find an approximation, when x is a small positive number, for $(1-x)^{-\frac{3}{4}}$, neglecting terms in x^3. Prove that the error is less than

$$35x^3/(16-18x).$$

11. Find an approximation, when x is a small positive number, for $\sqrt[3]{(1+x)}$, neglecting x^3, and prove that the error is less than $5x^3/(81-81x)$.

12. Write down series for (1) e^{-x}, (2) $\frac{1}{2}(e^x+e^{-x})$, (3) $\frac{1}{2}(e^x-e^{-x})$.

13. Give an approximation for e^{-x^2} as far as the term in x^4 and show that the error is numerically less than $2x^6/(12-3x^2)$.

14. Differentiate term by term the series for $\log_e(1+x)$ and find the sum of the resulting series.

15. Assuming the series for $\log_e(1+x)$ and $\log_e(1-x)$, prove that, for all positive values of y,

$$\log y = 2\left\{\frac{y-1}{y+1}+\frac{1}{3}\left(\frac{y-1}{y+1}\right)^3+\ldots\right\}.$$

16. What is the sum of the series $1-\frac{1}{2}+\frac{1}{3}-\frac{1}{4}+\ldots$?

17. Given $\log_e 10 = 2\cdot3026$, find $\log_e 11$ to 3 places of decimals.

18. Prove that $\log_e 1\cdot5$ is $0\cdot40546$, correct to 5 places of decimals.

19. Find the greatest term or terms in the expansion of e^5.

20. Prove that if $x - x^3/6$ is used as an approximation for $\sin x$, the error is numerically less than $7x^5/\{20\,(42 - x^2)\}$.

21. If $y = x/(1+x)$, where x is any positive number, show that $\log_e(1+x)$ may be expanded in ascending powers of y, and obtain the expansion. Deduce that for all positive values of x, $\log_e(1+x)$ is greater than $x/(1+x)$.

22. Show, by finding their values, that numbers a, b, c can be found such that the difference between the expansion of $e^x(1 - ax + bx^2 - cx^3)$ and $1 + ax + bx^2 + cx^3$ contains no power of x less than the seventh.

23. If the product of e^x and the polynomial $1 - x + x^2/2! - x^3/3!$ is written as a series in ascending powers of x, show that it contains no powers of x below the fourth, and find and simplify as much as possible the coefficient of x^n in the series, when n is not less than 4. Deduce that for all real values of x, e^{-x} is greater than $1 - x + x^2/2! - x^3/3!$.

24. Prove that, when x is sufficiently small, numerically,
$$(1+2x)^{-\frac{1}{2}}e^x = 1 + x^2 - \tfrac{4}{3}x^3 + \dots,$$
and find the coefficient of x^4.

25. Prove that, if x is so small that its fourth power may be neglected,
$$(1 - \tfrac{1}{2}x)^2(1+2x)^{\frac{1}{2}}/(1+3x^2)^{\frac{1}{2}} = 1 - \tfrac{9}{4}x^2 + \tfrac{5}{4}x^3.$$

26. Express $(3+4x)/\{(1+x)(1+2x)\}$ in partial fractions and hence obtain the first four terms of its expansion in powers of x. State the values of x for which the expansion is valid.

27. Expand $(5-4x)/\{(1-x)(2-x)\}$ as far as the term in x^4, stating the range for which the expansion is valid. Give also the term in x^n.

28. Show that, when x is small, $x^2/\{(1-x)^2(1-2x)\}$ is approximately equal to $x^2 + 4x^3$.

29. Express $(2x+1)/\{(x-2)(x+1)^2\}$ in the form
$$A/(x-2) + B/(x+1) + C/(x+1)^2,$$
where A, B, C are independent of x. Deduce that, when the given expression is expanded in ascending powers of x, the coefficient of x^n is
$$-\tfrac{1}{9}\{5/2^{n+1} \pm (3n-2)\},$$
the plus or minus sign being taken according as n is odd or even.

30. Find the values of b and c in terms of a when the expansion of e^{ax} in powers of x is identical with the expansion of $(1+bx)/(1-cx)$ as far as the term in x^2, assuming that a is not zero. Deduce that with these values of b and c, $(1+bx)/(1-cx) - e^{ax} = \frac{1}{12}a^3x^3$, when x is so small that x^4 can be neglected.

31. Write down the expansion of $\log_e\{(1+x)/(1-x)\}$ in ascending powers of x, stating the limits of x for which the expansion is valid. Expand $\log_e(1+1/n)$ in ascending powers of $1/(2n+1)$, stating the necessary restrictions on the value of n. Prove that if n has any positive value, $\log_e(1+1/n)$ is greater than $2/(2n+1)$ and less than

$$(2n+1)/\{2n(n+1)\}.$$

32. Show that if $\log_e y = 1 + \frac{1}{2}x - \frac{1}{6}x^2 + \frac{1}{12}x^3$, then as far as the third power of x, $y = e(1 + \frac{1}{2}x - \frac{1}{24}x^2 + \frac{1}{48}x^3)$.

33. Show that, if $n > 1$,

$$\log_e\{(n+1)/(n-1)\} = 2\{1/n + 1/(3n^3) + 1/(5n^5) + \ldots\},$$

and calculate $\log_e 9$ (as $2\log_e 3$) correct to 4 places of decimals.

34. The volume of a cone is given by $V = \frac{1}{3}\pi r^2 h$, where h is the height and r is the radius of the base. Express r in terms of V and h, and find approximately by what fraction r must be increased if the volume is to remain unaltered while the height is diminished by a small amount x.

35. If T is the periodic time of a planet and a is the major semi-axis of its orbit, Kepler's third law states that T^2 varies as a^3. If a is increased by a small amount x per cent, prove that the corresponding small increase in T is approximately $1\cdot5x$ per cent.

36. A sports track 4 yards wide begins with a straight section of 150 yards, after which it curves to the left. If a runner who starts on the extreme right runs in a straight line towards the corner, how many inches farther has he to go than one starting on the extreme left?

37. The time of swing of a pendulum is given by the formula

$$T = 2\pi\sqrt{(l/g)},$$

where l is the length. Show that if the pendulum of a clock which normally beats seconds is increased in length by $\frac{1}{10}$ per cent, the clock will lose approximately $43\frac{1}{4}$ seconds per day.

38. During a certain period of the day, the cashiers of a department store have to deal with an average of 30 transactions a minute. The probabilities that during a given minute there should be 0, 1, 2, 3, ... transactions are $1/e^{30}$, $30/(1!\,e^{30})$, $30^2/(2!\,e^{30})$, $30^3/(3!\,e^{30})$, etc. Show that these probabilities increase up to 29 transactions, that their total, if the series is continued indefinitely, is 1, and find the chance that during a given minute the number of transactions should be exactly 50.

$$(\log_{10} 50! = 64\cdot4831.)$$

39. At a certain period of the day a telephone exchange has to deal on the average with 10 calls per minute. The chances that during a given minute 0, 1, 2, 3,... calls will occur are represented by the terms of the series $\sum_{r=0}^{\infty} \{(10^r/r!)\,e^{-10}\}$. What is the sum of this series? What is the chance that during a given minute there will be (1) exactly 4 calls, (2) less than 4 calls, (3) more than 4 calls?

40. If in the above question we multiply each possible number of calls by the probability of that number occurring, and add the results, i.e. if we form the series $\sum_{r=0}^{\infty} \{r\,(10^r/r!)\,e^{-10}\}$, we obtain the 'expected' number of calls per minute. Show that this is 10.

41. Write down, for the above telephone exchange, the terms of the series giving the probabilities of 21, 22, ... calls in a given minute, and show by a comparison with a geometric series that the chance of their being more than 20 calls is less than 0·002.

$$(\log_{10} 21! = 19\cdot708.)$$

42. Simplify the following expression, which occurs in the application of the Quantum Theory to the Kinetic Theory of Gases:

$$\frac{\epsilon x + 2\epsilon x^2 + 3\epsilon x^3 + ...}{1 + x + x^2 + x^3 + ...},$$

x being a small quantity, equal to $e^{-\epsilon/kT}$.

43. In the Kinetic Theory of Gases, the mean energy of a molecule for each degree of freedom is $kT/2$. Under Quantum Theory assumptions this is replaced by the corrected formula $\tfrac{1}{2}h\nu/(e^{h\nu/kT} - 1)$. Show that if T is so large that $e^{h\nu/kT}$ may be replaced by the first two terms of its expansion, the corrected formula reduces to the original one.

44. The 'dip of the horizon', $\varDelta$, for an observer at a height h above sea level is given by $\cos \varDelta = R/(R+h)$, R being the earth's radius. Prove that, approximately, $\varDelta = \sqrt{(2h/R)}$.

45. The length of a degree of latitude, at a place whose latitude is ϕ, is given, allowing for the ellipticity of the earth, by the formula

$$68 \cdot 7 / \{1 - \tfrac{26}{3963} \sin^2 \phi\}^{\frac{3}{2}} \text{ miles.}$$

Show that this can be written approximately as

$$68 \cdot 7 \{1 + 0 \cdot 00984 \sin^2 \phi\} \text{ miles.}$$

X. STATISTICS

10.1. Introduction.

The original meaning of 'Statistics' was simply 'the description of states'; the term was used for any information, not necessarily numerical, about countries and their inhabitants. The 'political arithmetic' which began in England in 1662 with Graunt's *Natural and Political Observations upon the Bills of Mortality*, was much nearer to the modern meaning of 'statistics'. Not until after 1800, when the results of the theory of probability began to be applied in statistical work, did the mathematical side of the subject develop. Quetelet, a brilliant Belgian writer, famous for his theory of 'l'homme moyen', did much to bring statistics before the public eye, and between 1853 and 1876 there was a series of nine international statistical congresses. In the meantime, actuarial science had developed considerably, and the study of vital statistics led to various theoretical problems in which the Germans, Knapp, Zeuner and Lexis, and the English actuary Woolhouse, were prominent.

Statistical method was first applied to the study of heredity by Galton in 1869. He emphasised the idea of deviations from an average, and in his *Natural Inheritance*, published 20 years later, the idea of 'regression' is put forward. By this word he meant the tendency for exceptional parents to have children nearer the mean than themselves.

This idea, and that of correlation, were developed by Karl Pearson and Yule. The word 'regression' lost its original meaning and was applied to any relation expressing the mean value of a variable in terms of a given value of a correlated variable. The application of statistical method to biological problems was further advanced by Fisher.

Biology is now an experimental science, but in experiments on living organisms the conditions are less easily controlled than in physics and chemistry. Hence the experiments must be repeated many times over and the results analysed. Therein lies the importance of statistics and probability to the biologist.

10.2. Statistics.

A statistic is a number, such as, for example, a mean, which is taken as in some way characteristic of a large group of numbers. The mean, or average, is indeed the only statistic which is commonly used for everyday affairs, and but little reflection shows that taken by itself it gives a very inadequate representation of the group of numbers which it is taken to represent. To say, for example, that a batsman's average for the season is 10, leaves a good deal in doubt as regards his actual performance.

10.3. Frequency distributions.

Where a large collection of observations is involved it is usual to divide them into groups, each group covering the same range of the quantity observed, and to record, or plot on a graph, the number of observations in each group. The following table,* for example, gives the results of measurements made on 209 spikes (heads) of a variety of barley to determine the distance apart of the spikelets (flowers):

Dist. in mm.	3·50–3·69	3·70–3·89	3·90–4·09	4·10–4·29	4·30–4·49	4·50–4·69
Number in class	0	1	3	26	63	60

Dist. in mm.	4·70–4·89	4·90–5·09	5·10–5·29	5·30–5·49	5·50–5·69
Number in class	36	15	4	0	1

The figures may also be exhibited graphically:

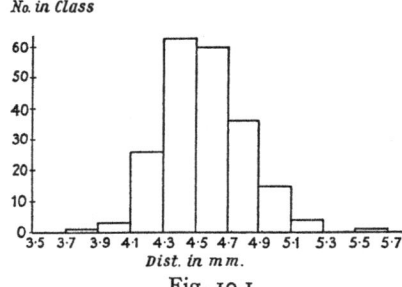

Fig. 10.1

This, like many other examples of frequency distribution graph, bears some resemblance to the shape of the graphs shown in Fig. 6.1. It may easily be imagined that if more of these curves had been drawn, for increasing values of n, and if the scale had been continually decreased, the shape of the curves would have approximated more and more closely to that shown below.

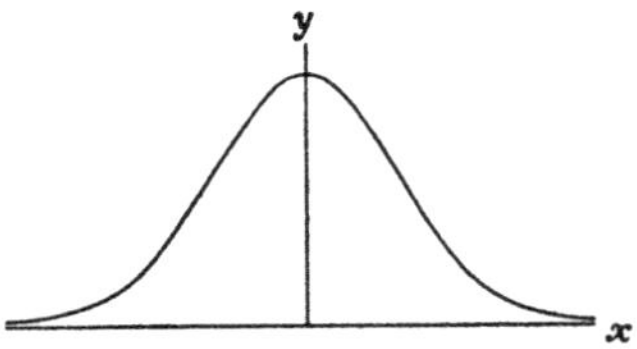

Fig. 10.2

* Reprinted, by permission of the McGraw Hill Publishing Co. Ltd., from Hayes and Garber's *Breeding Crop Plants*.

This is called the 'Normal curve of errors' and its equation, if suitable scales are used, can be taken as $y = e^{-x^2}$ (or better, $y = \dfrac{1}{\sqrt{(2\pi)}}\, e^{-x^2/2}$—the area under the curve is then one unit). Not all frequency distributions are of this type, however. Some are unsymmetrical, resembling the probability distributions mentioned in §7.6; some bear a close relation to the terms of an exponential series, rather than to those of a binomial expansion (see Ex. IX, 38–41); and some are entirely irregular. But it is often useful to relate them in suitable cases to the model distributions mentioned, and for this purpose it is obviously necessary to have certain measures of their character. Of these the mean may be the first; it gives us a value which may be taken as the centre of the distribution. But we shall also need to measure the degree of spread away from that centre, and the degree of 'skewness' or lack of symmetry. We have to distinguish, in fact, between the shapes shown in Fig. 10.3, all of which represent distributions of the same number of observations, having the same mean.

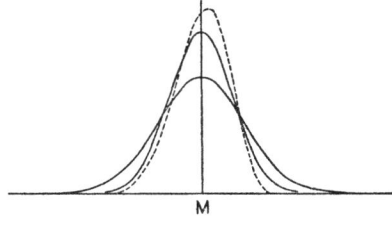

Fig. 10.3

It should be observed that as the number of observations is the same, the areas under the curves are equal; and as the means are all the same, the centres of gravity of those areas are all vertically above the point M on the axis.

10.4. Mode.

There is no one method of measuring these characteristics. For example, it may be argued that in the case of an unsymmetrical distribution, the mean is not the best point to take as a centre. It might be considered more natural to take the value at the point where the curve is highest. This value is called the 'mode' (it is the most frequently occurring or most fashionable value). In the example given in §10.3 this may be taken very roughly as 4·4, the centre of the biggest group.

10.5. Median and quartiles.

Another value which has a good deal to recommend it is the 'median': this is the value which would be the middle one if the results of all the observations were arranged in order of size. In the example given, as there are 209 observations recorded, the median will be the one which is No. 105 from either end. This will be a member of the group 4·5–4·69, No. 12 from the 4·5 end, and if it is supposed that the 60 members of the group are uniformly distributed, its value will be 4·54 mm.

The spread may be measured on the same principle by means of the 'quartiles': if the whole range of observations, arranged in order, is divided into 4 equal parts, the boundary values are respectively the 'lower quartile', the 'median' and the 'upper quartile'. If we divide 210 (1 more than the total number of observations) by 4, the result is $52\frac{1}{2}$. The three boundary values required are those corresponding to observations No. $52\frac{1}{2}$, No. 105 and No. $157\frac{1}{2}$. The lower quartile is therefore in the group 4·30–4·49, and its value is $4\cdot30 + \frac{22\frac{1}{2}}{63} \times 0\cdot2$, i.e. 4·37 mm. The upper quartile is similarly $4\cdot70 + \frac{4\frac{1}{2}}{36} \times 0\cdot2$, i.e. 4·72 mm. The 'quartile deviation', which is a measure of the spread, is half the difference between the upper and lower quartiles, i.e. $\frac{1}{2}(4\cdot72-4\cdot37)$, $= 0\cdot175$ mm.

10.6. Mean and standard deviation.

The easiest method of calculating the mean is to adopt a change of scale, re-numbering the classes so that the larger frequencies occur in the classes with small numbers. For example we may re-label the frequency diagram as follows:

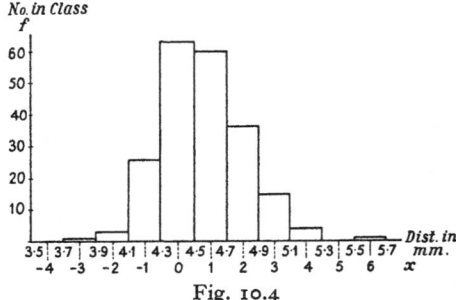

Fig. 10.4

In practice we work in tabular form, as follows:

Dist. in mm.	x	f	fx	
			$+$	$-$
3·50–3·69	−4	0		
3·70–3·89	−3	1		− 3
3·90–4·09	−2	3		− 6
4·10–4·29	−1	26		−26
4·30–4·49	0	63		
4·50–4·69	1	60	60	
4·70–4·89	2	36	72	
4·90–5·09	3	15	45	
5·10–5·29	4	4	16	
5·30–5·49	5	0		
5·50–5·69	6	1	6	
	Totals	209	199	− 35
				= 164

The column headed fx is obtained by multiplying those headed f and x. Positive and negative values are separated for convenience. The total of this column, 164, is the sum of all the 209 values of x, and may be called Σx. (Some writers call it Σfx, but it should be realised that the grouping is not an essential feature of the process. There are 209 values of x, and Σx is their sum.)

The mean value of x, which we shall call $\bar{x}$, is given by

$$\bar{x} = \frac{\Sigma x}{n}, \text{ where } n = \text{total no. of observations,}$$

$$= \tfrac{164}{209} = 0 \cdot 785.$$

Changing back to the original scale, the mean value is therefore

$$4 \cdot 4 + 0 \cdot 785 \times 0 \cdot 2 = 4 \cdot 56 \text{ mm.}$$

To obtain a corresponding measure of the spread of the values, we must naturally consider deviations from the mean. Let y be the value given by any observation, measured from the mean as origin, the scale being otherwise the same as for x. In short let

$$x = \bar{x} + y. \tag{1}$$

The values of y will be some positive and some negative, and since y is measured from the mean their total will be zero. (Algebraically, adding up all the n equations of type (1), we obtain

$$\Sigma x = n\bar{x} + \Sigma y.$$

But $$\bar{x} = (\Sigma x)/n.$$

Therefore $$\Sigma y = 0.)$$

The sum of all the deviations from the mean is zero, but the sum of their squares will be necessarily positive, and will give us a measure of the spread of the figures. To calculate this, we must first observe that

$$x^2 = (\bar{x} + y)^2 = \bar{x}^2 + 2\bar{x}y + y^2.$$

Therefore $$\Sigma x^2 = n\bar{x}^2 + 2\bar{x}.\Sigma y + \Sigma y^2.$$

But $$\Sigma y = 0.$$

Therefore $$\Sigma x^2 = n\bar{x}^2 + \Sigma y^2$$

and $$\Sigma y^2 = \Sigma x^2 - n\bar{x}^2.$$

Therefore the 'mean square deviation' or 'variance', namely*

$$\frac{\Sigma y^2}{n} = \frac{\Sigma x^2}{n} - \bar{x}^2, \tag{2}$$

and its value can be calculated by continuing the above table as follows:

Dist. in mm.	x	f	fx +	fx −	fx^2 +
3·50–3·69	−4	0			
3·70–3·89	−3	1		− 3	9
3·90–4·09	−2	3		− 6	12
4·10–4·29	−1	26		−26	26
4·30–4·49	0	63			
4·50–4·69	1	60	60		60
4·70–4·89	2	36	72		144
4·90–5·09	3	15	45		135
5·10–5·29	4	4	16		64
5·30–5·49	5	0			
5·50–5·69	6	1	6		36
	Totals	209	199 = 164	−35	486

As before, $$\bar{x} = \frac{\Sigma x}{n} = \frac{164}{209} = 0.7847.$$

* In dealing with small samples from a large 'population' it is better to divide by $(n-1)$ rather than n.

Mean square deviation (or 'variance')

$$= \frac{\Sigma x^2}{n} - \bar{x}^2$$

$$= \tfrac{486}{209} - (0\cdot7847)^2$$

$$= 2\cdot3254 - 0\cdot6158$$

$$= 1\cdot7096.$$

'Standard deviation' $= \sqrt{(\text{mean square deviation})}$

$$= 1\cdot308.$$

Reverting to the original scale, the standard deviation, usually called σ,
$= 0\cdot262$ mm.

The standard deviation is the most useful measure of the spread of the figures. For a 'normal' distribution the quartile deviation is approximately $\tfrac{2}{3}\sigma$. For the example given, $\tfrac{2}{3}\sigma = 0\cdot175$ mm., and the quartile deviation was $0\cdot175$ mm., an unusually close agreement.

For a 'normal curve' the positions of the mean, median and quartiles and the magnitude of the standard deviation are as shown below:

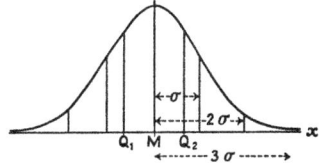

Fig. 10.5

The median and mean coincide at M.

(The analogy in mechanics of standard deviation is radius of gyration about an axis through the centre of gravity. It will be recalled that the square of this is less than that about any parallel axis by an amount equal to the square of the distance between the two, a result corresponding exactly to equation (2) above.)

10.61. Coefficient of variation.

It is sometimes useful to compare the spread with the mean value. For this purpose it is convenient to use the 'coefficient of variation' or 'coefficient of variability', which is the standard deviation expressed as a percentage of the mean, i.e. $100\sigma/\text{mean}$.

In the example given, this is $\dfrac{100 \times 0\cdot262}{4\cdot56}$, i.e. $5\cdot7$.

10.62. Probable error.

Another way of indicating the spread of the figures is to give the 'probable error', which, like the quartile deviation, is about ⅔ of the standard deviation. Thus we might say that the distance between the barley spikelets is 4·56 mm., with a p.e. of 0·17 mm., or we might say that the distance between the barley spikelets is 4·57 ± 0·17 mm. It will be seen from this that 'probable error' means the variation, or error, that is as likely as not to be exceeded.

'Probable error', however, is not much used, because no one is interested in a result which has a 50 per cent chance of being right. It is more usual to give σ, the 'standard error' (i.e. standard deviation), and to bear in mind that for a 'normal' distribution, the odds against an error of more than 2σ are about 21 to 1. Similarly the odds against an error of more than 3σ are about 370 to 1. (It will be noticed that in Fig. 10.5 only a very small fraction of the area under the curve lies outside the range -2σ to $+2\sigma$.)

10.63. Standard errors of mean and standard deviation.

If another set of barley heads were measured, it is not to be supposed that they would give exactly the same values as before for the mean and the standard deviation and other statistics. We must in fact expect a variation in these results due to chance, especially if the samples which we work with are small. Hence we speak of the 'standard error of the mean'. It is to be expected that it will be greater as the sample is smaller. It can be proved that for a 'normal' distribution it is approximately $\dfrac{\sigma}{\sqrt{n}}$, n being the number of observations in the sample.

Similarly it can be proved that the S.D. (standard deviation), σ, is subject to a standard error of approximately $\dfrac{\sigma}{\sqrt{(2n)}}$.

In the example given above, the mean is 4·56, with a standard error of 0·018, and $\sigma = 0·262$, with a standard error of 0·013.

10.7. Linear regression.

The following table and graph show the results of an experiment to determine the absorption of oxygen by the lungs as the work done by the body increases:

| x (work done) | 300 | 420 | 680 | 840 | 954 | Kg. metres |
| y (oxygen absorbed) | 680 | 820 | 1000 | 1080 | 1300 | c.c. per min. |

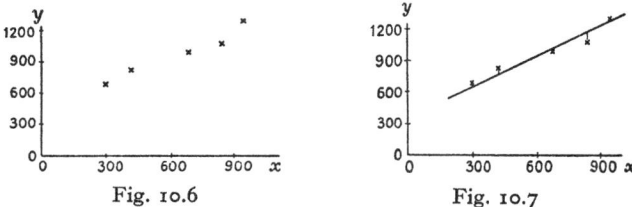

Fig. 10.6   Fig. 10.7

It appears at once that a straight line could be drawn passing not far from any of the points. Any such line drawn to represent the trend of the figures is called a 'regression line', and if, as in this case, it is straight, we say that the regression is 'linear'. We might choose the exact position of the line by eye, or we might adopt the more precise method given below. The equation of the line can be found by the methods explained in Chapter II, and is called a 'regression equation'. (As already stated in § 10.1, the word 'regression' no longer means anything in the nature of a falling back: we are simply representing the trend of the figures by an equation for y in terms of x.)

There is clearly a wide choice as to exactly where the line should be drawn. If we are proposing to express y in terms of x, the deviations of the points from the line are measured in the y-direction (see Fig. 10.7) and our object is to make these as small as possible. It is usually considered best to draw the line in such a way that the sum of the squares of the deviations is a minimum. This is known as the 'method of least squares'. It can be proved that in order to do this the line must be drawn through the 'mean centre' $(\bar{x}, \bar{y})$ of the n points $\left(\text{i.e. the point given by } \bar{x} = \dfrac{\Sigma x}{n}, \bar{y} = \dfrac{\Sigma y}{n} \right)$ and that its gradient must be p/σ_x^2, where $p = \dfrac{1}{n}\Sigma\xi\eta$, and $\sigma_x^2 = \dfrac{1}{n}\Sigma\xi^2$, (ξ, η) being the coordinates of one of the n points referred to the centre of gravity as origin.

10.71. Proof.

If the equation of the line is $y = mx + c$ and the given points are $(x_1 y_1), (x_2 y_2)$, etc., the deviations are $y_1 - mx_1 - c, y_2 - mx_2 - c$, etc.

Let $S = \Sigma(y_1 - mx_1 - c)^2$.

We have to choose m and c in such a way as to make this quantity a minimum.

LA   6

First consider m as a variable, all the other letters being constants.

Differentiating, $\dfrac{dS}{dm} = \Sigma 2(y_1 - mx_1 - c)(-x_1).$

For a minimum, this must be zero.

Therefore $\quad\quad \Sigma x_1 y_1 - m\Sigma x_1{}^2 - nc = 0.$ (1)

Now consider c as a variable, all the other letters being constants.

$$\frac{dS}{dc} = \Sigma 2(y_1 - mx_1 - c)(-1).$$

For a minimum, this also must be zero.

Therefore $\quad\quad \Sigma y_1 - m\Sigma x_1 - nc = 0.$ (2)

Equations (1) and (2) determine the required values of m and c.

Equation (2) may be written

$$n\bar{y} - mn\bar{x} - nc = 0.$$

Therefore $\quad\quad \bar{y} - m\bar{x} - c = 0.$

If the origin is taken at the mean centre, $\bar{x}$ and $\bar{y}$ are both zero, and hence $c = 0$. Equation (1) then shows that

$$m = \Sigma \xi\eta / \Sigma \xi^2$$
$$= p/\sigma_x{}^2.$$

10.72. Application.

Using the figures given in § 10.7, we may construct the following table:

	x	y	ξ $(x-637)$	η $(y-976)$	$\xi\eta$ (nearest 100)	ξ^2 (nearest 100)	η^2 (nearest 100)
	300	680	-337	-296	99,800	113,600	87,600
	420	820	-217	-156	33,900	47,100	24,300
	680	1,000	$+43$	$+24$	1,000	1,800	600
	840	1,080	$+203$	$+104$	21,100	41,200	10,800
	945	1,300	$+308$	$+324$	99,800	94,900	105,000
Total	3,185	4,880	0	0	255,600	298,600	228,300
Total divided by 5	637	976			51,120	59,720	45,660

$(\bar{x}, \bar{y})$ is (637, 976),

and $m = p/\sigma_x{}^2,\ = 51{,}120/59{,}720,\ = 0{\cdot}856.$

Using Method 2 of §2.3, the regression equation is
$$y - 976 = 0.856\,(x - 637),$$
or
$$y = 0.856\,x + 431. \tag{3}$$

In practice, it is often quicker not to tabulate ξ and η, but to use the facts (proved in §10.82) that

$$p = \frac{\Sigma\xi\eta}{n} = \frac{\Sigma xy}{n} - \overline{xy}, \quad \text{and} \quad \sigma_x^2 = \frac{\Sigma x^2}{n} - \bar{x}^2. \tag{4}$$

The number 0.856 (equal to p/σ_x^2) is called the 'regression coefficient' of y on x.

In the same way we could express by another equation the regression of x on y. The corresponding coefficient would be p/σ_y^2, and the regression equation would be
$$x - \bar{x} = p/\sigma_y^2(y - \bar{y}),$$
or in the case given,
$$x - 637 = 1.12\,(y - 976),$$
i.e.
$$x = 1.12y - 456. \tag{5}$$

The reason why this result is different from equation (3) is that we are here concerned with deviations measured parallel to the x-axis. The formula gives, in fact, a mean value of x for a given value of y, instead of a mean value of y for a given x.

10.8. Correlation.

The following table shows the marks of 35 candidates in two mathematical papers in a scholarship examination:

Paper		Paper		Paper	
I	II	I	II	I	II
20	12	8	0	2	20
18	31	15	1	54	33
22	10	24	38	55	24
46	40	35	9	56	22
6	5	31	28	12	26
0	0	29	62	39	47
19	4	43	31	25	5
57	37	34	41	63	56
46	56	34	36	27	37
31	3	48	19	27	16
45	63	28	13	23	9
40	27	46	66		

A first glance at these figures shows that on the whole the candidates who did badly at one paper were weak on the other also. Others did well at both. But there are many exceptions, and it is impossible to assess the closeness of the agreement without careful analysis. The object of this part of the chapter is to show how a numerical measure can be given to the degree of 'correlation', as it is called, between the two sets of figures.

Representing the marks on the first paper by x, the second by y, we may plot a diagram as follows:

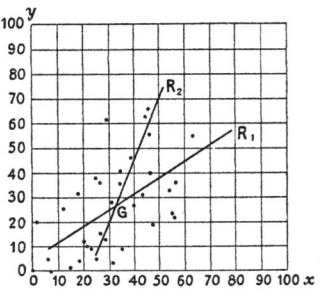

Fig. 10.8

Each dot represents a pair of values, and it will be observed that the area covered by the dots is very roughly an ellipse, its major axis lying in a direction between that of Ox and Oy. The closer the correlation between the two sets of figures, the more does this ellipse narrow. If all the dots were to lie on a straight line, that would illustrate what is regarded as perfect correlation, the differences in the values of x being proportional to those of y. (It should be noted that perfect agreement between the two sets of marks does not imply that the corresponding values of x and y should be equal, or even proportional; it is only necessary that the *differences* between values of x should be proportional to the corresponding differences for y.) If on the other hand there were no relation at all between the two sets of values, the diagram would assume something like a circular form, or an ellipse with its axes parallel to those of x and y, and if there were 'negative correlation' (i.e. if an increase in x corresponded, as a rule, to a decrease in y) the major axis of the ellipse would slope 'downwards' (i.e would make a negative angle with Ox).

The line GR_1 is drawn in such a way that the sum of the squares of the vertical distances of the dots from the line is a minimum; and GR_2 is drawn so that the sum of the squares of the horizontal distances of the dots from the line is a minimum. These are the 'regression lines'. It is evident that the greater the degree of correlation the more nearly will they coincide, and that if there is no correlation, they will be at right angles. It has been proved (in §10.71) that G, their point of intersection, is the mean centre of the system of dots; that the gradient of GR_1 is p/σ_x^2, and that of GR_2, measured with respect to Oy, is p/σ_y^2; σ_x, σ_y being the standard deviations of the two sets of values, and p being $(\Sigma\xi\eta)/n$, when (ξ, η) are the coordinates of any point referred to axes through G parallel to Ox and Oy.

10.81. Coefficient of correlation.

The value of $p/\sigma_x\sigma_y$ is called the 'coefficient of correlation'. For perfect correlation, the lines coincide, the aforementioned gradients are reciprocals, and $p/\sigma_x\sigma_y = 1$; when there is no correlation, the lines are parallel to the axes and $p/\sigma_x\sigma_y = 0$; and for perfect negative correlation the lines again coincide, and $p/\sigma_x\sigma_y = -1$. (N.B. σ_x and σ_y are necessarily positive, but p may be positive or negative.) The coefficient of correlation, usually denoted by r, must lie between -1 and $+1$.

10.82. Calculation of r.

It is usually convenient to group the figures, so as to avoid the labour of evaluating ξ^2, $\xi\eta$ and η^2 for each individual pair of values.

The next figure shows the readings of §10.8 grouped in this way.

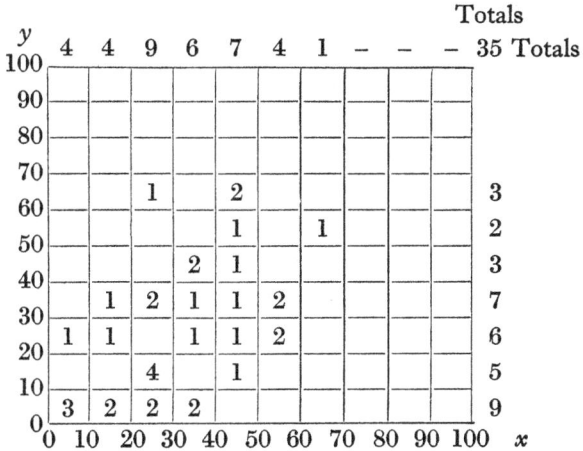

As before, it is advisable to adopt new scales, as indicated in the table below. The means $\overline{X}, \overline{Y}$ and the standard deviations σ_x, σ_y are worked out in the same manner as before. To evaluate the product terms ΣXY, it is convenient to write in the corner of each square the value of XY. ΣXY can then be quickly found. But

$$\Sigma XY = \Sigma(\overline{X}+\xi)(\overline{Y}+\eta),$$

where $(\overline{X}, \overline{Y})$ are the coordinates of G,

$$= \Sigma\overline{XY}+\overline{X}\Sigma\eta+\overline{Y}\Sigma\xi+\Sigma\xi\eta$$
$$= n\overline{XY}+\Sigma\xi\eta,$$

since $\Sigma\eta$ and $\Sigma\xi$ are zero (as in the table of §10.72).

Therefore $\qquad \Sigma\xi\eta = \Sigma XY - n\overline{XY}$

and $\qquad p = (\Sigma XY)/n - \overline{XY}.$

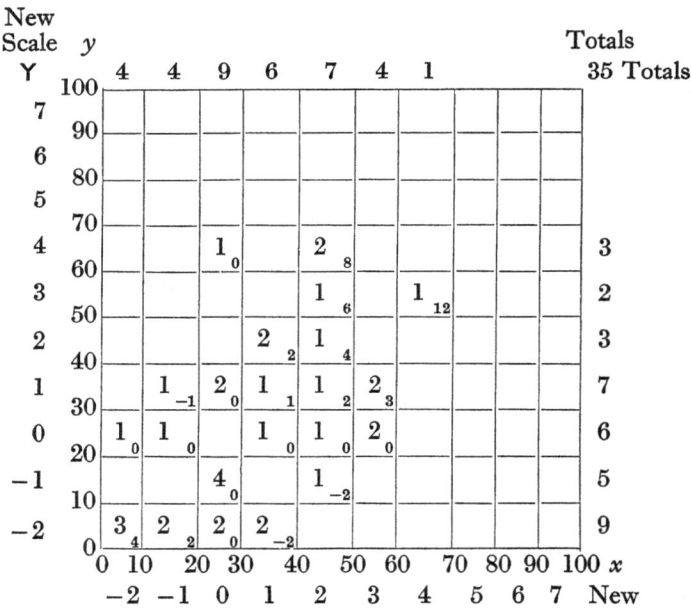

X	f	fX	fX²	Y	f	fY	fY²	fXY +	fXY −
4	1	4	16	4	3	12	48	16	
								6	
3	4	12	36	3	2	6	18	12	
								4	
2	7	14	28	2	3	6	12	4	
									−1
1	6	6	6	1	7	7	7	1	
								2	
0	9	.	.	0	6	.	.	6	
									−2
−1	4	−4	4	−1	5	−5	5	12	
								4	
−2	4	−8	16	−2	9	−18	36		−4
Totals	35	24	106		35	8	126	67	−7
									= 60

Divide by 35: 0·69, 3·03,   0·23, 3·60,   1·71.

Therefore $\overline{X} = 0\cdot69$. $\overline{Y} = 0\cdot23$.

$$p = 1\cdot71 - 0\cdot69 \times 0\cdot23 = 1\cdot55.$$

$$\sigma_x = \sqrt{3\cdot03 - (0\cdot69)^2} = 1\cdot60. \qquad \sigma_y = \sqrt{3\cdot60 - (0\cdot23)^2} = 1\cdot88.$$

$$r = \frac{1\cdot55}{1\cdot60 \times 1\cdot88} = 0\cdot52.$$

10.83. Significance of values of r.

The significance of any particular value of r depends on the number of pairs of readings from which it has been obtained. If only a few pairs of readings have been used, quite a large value of r might be obtained as a result of chance selection, even though there might be no correlation at all in the complete 'population' from which a sample has been taken. Tables are given in Fisher's *Statistical Methods for Research Workers* showing the probability of any given value being obtained through the element of chance introduced by sampling. The diagram, based on these tables, shows the values of which there is a 1 per cent probability for different values of n.

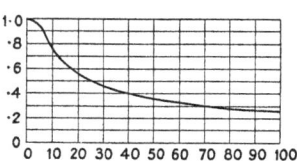

Fig. 10.9

It will be seen that if there were no correlation, the probability of obtaining the value 0·52 from a sample of 35 pairs of readings would be less than 1 per cent.

EXAMPLES X

1. The following table gives the marks of a batch of 83 candidates in School Certificate Elementary Mathematics.

Marks	No. of candidates	Marks	No. of candidates
70– 79	1	190–199	4
80– 89	3	200–209	7
90– 99	1	210–219	10
100–109	0	220–229	3
110–119	1	230–239	3
120–129	7	240–249	5
130–139	2	250–259	3
140–149	5	260–269	7
150–159	1	270–279	2
160–169	8	280–289	1
170–179	3	290–299	1
180–189	5		

Find the mean and the standard deviation. Compare $\frac{2}{3}\sigma$ with the quartile deviation.

2. The weights in grammes of 35 Light Sussex - Rhode Island Red cockerels were as follows:

2840	2430	2695	2315	2545
2815	2365	2655	1855	2530
2685	2320	2575	3300	2410
2680	2210	2545	2795	2385
2560	1755	2505	2760	2295
2555	2930	2470	2625	2150
2475	2795	2365	2595	2060

Divide them into groups at intervals of 100 gm., and draw a frequency diagram. Work out the mean and the S.D. and draw lines on the diagram to indicate the mean and the 'standard error'. Find also the median and quartiles, marking them on the diagram, and the quartile deviation.

3. The following table shows the distribution of Intelligence Quotients among 732 children of both sexes.

Value of I.Q.	No. of boys	No. of girls
109	0	0
108	1	0
107	1	0
106	1	0
105	0	0
104	0	3
103	5	1
102	1	1
101	7	1
100	8	5
99	6	4
98	10	8
97	6	5
96	13	6
95	11	9
94	6	7
93	7	13
92	18	13
91	12	11
90	18	21
89	14	14
88	19	20
87	19	18
86	18	21
85	19	22
84	19	20
83	14	13
82	18	17
81	10	18
80	14	16
79	11	6
78	10	12
77	8	6
76	14	8
75	5	12
74	8	8
73	5	11
72	6	5
71	5	6
70	3	1
Totals	370	362

Grouping the marks 70–74, 75–79, etc., find the mean and S.D., (a) for the boys, (b) for the girls. Also the corresponding quartile deviations.

4. The weight of a side of bacon when cured bears a definite relation to the weight of the live pig before it is killed. The following table gives figures for Berkshire pigs:

Live weight of pig	124	152	187	201	219	lb.
Weight of cured side of bacon	30	40	50	60	70	lb.

Find an equation for the regression of the weight of the side of bacon on that of the live pig.

5. The following table* gives indices of employment and business activity between the years 1929 and 1937, the figures for 1935 being taken as a standard in each case:

	1929	1930	1931	1932	1933	1934	1935	1936	1937
Business activity	$98\frac{1}{2}$	93	$87\frac{1}{2}$	84	89	96	100	106	112
Employment	$98\frac{1}{2}$	$94\frac{1}{2}$	91	90	$93\frac{1}{2}$	$97\frac{1}{2}$	100	105	111

Tabulate the deviations from 100, and use the equations (4) of § 10.72 to obtain an equation for the regression of employment on business activity.

6. The following table shows the marks of 35 candidates in two papers, one a Latin Unseen, the other Mathematics, in a scholarship examination.

L	M	L	M	L	M
51	12	19	27	4	66
67	31	6	0	43	20
39	10	73	1	44	33
57	40	32	38	48	24
16	5	71	9	79	22
0	0	82	28	45	26
0	4	67	62	60	47
46	37	45	31	0	5
66	56	41	41	28	56
71	3	53	36	14	37
76	63	76	19	46	16
		79	13	0	9

Find the correlation coefficient.

* Reprinted by permission from *The Economist Trade Supplement*, 25 February 1939.

7. 21 candidates were examined in Latin, English and Mathematics, the marks being as shown below:

L	E	M	L	E	M
108	48	136	128	64	45
93	46	28	100	28	16
66	48	32	168	67	40
30	29	15	112	44	84
35	26	38	162	50	52
87	43	26	142	71	117
55	31	54	131	51	92
24	31	4	138	34	87
54	28	34	34	33	81
128	65	130	140	43	114
65	38	25			

Find the correlation coefficients for each pair of subjects.

8. Flour from various strains of wheat was milled and baked under uniform controlled conditions, the object being to determine whether there was any correlation between the percentage of protein in the various flours, and the size of the loaf when baked. The results of the various trials are given here. What is the correlation coefficient?

Loaf volume in c.c.	Protein content %	Loaf volume in c.c.	Protein content %
1980	14·9	1990	13·4
2030	14·3	2010	14·1
2235	17·1	2060	14·1
2245	15·4	2000	14·1
2285	15·3	1930	14·9
2225	14·7	1970	13·7
2030	14·3	1980	13·7
2070	14·7	2010	13·6
2010	14·1	2020	13·4
2000	13·9	2010	13·9

(The figures may conveniently be divided into groups covering 20 c.c. and 0·2 %.)

9. The following table shows the annual rainfall and sunshine recorded at Felsted from 1912 to 1937:

	Rainfall in inches	Sunshine in hours		Rainfall in inches	Sunshine in hours
1912	26·37	1459	1925	24·83	1602
1913	18·86	1454	1926	24·03	1374
1914	24·71	1693	1927	29·21	1480
1915	25·49	1675	1928	25·33	1665
1916	31·26	1468	1929	20·51	1712
1917	23·67	1692	1930	23·80	1517
1918	25·76	1659	1931	23·32	1329
1919	24·71	—	1932	20·12	1371
1920	24·00	1663	1933	15·23	1605
1921	14·41	1931	1934	19·40	1468
1922	21·30	1606	1935	26·55	1544
1923	22·74	1519	1936	23·19	1155
1924	31·99	1618	1937	27·75	1314

Neglecting the year 1919, find the correlation coefficient between the rainfall and the sunshine. (Take groups covering 1 inch and 100 hours.)

10. The following table* gives a comparison of the fat content and moisture content of the carcases of 40 Light Sussex cockerels:

% Moisture	% Fat	% Moisture	% Fat
69·07	5·47	68·25	7·45
70·15	5·93	70·52	6·20
68·01	7·27	69·90	7·10
73·30	2·72	68·61	8·02
70·42	5·43	65·80	11·31
69·18	8·02	69·66	6·31
67·46	8·78	69·60	5·54
68·87	7·83	71·99	4·12
67·69	7·82	69·13	7·01
65·64	9·40	70·25	4·27
72·90	2·88	67·17	7·40
69·50	6·28	68·53	7·60
69·63	6·72	69·90	6·52
69·55	6·50	67·54	7·80
70·50	5·83	66·62	8·62
70·51	6·44	67·28	9·02
71·22	4·94	70·91	5·21
64·74	12·04	67·36	7·96
64·94	11·29	65·94	10·55
66·56	10·13	66·42	9·92

Find regression equations and the correlation coefficient. (Take covering groups 1 %.)

* Reprinted by permission of the editor of the *Journal of Agricultural Science* from 'Some Observations on the Normal Variations in Composition of Light Sussex Cockerels' by E. T. Halnan.

11. A rough measure of correlation can be obtained by the following rule of Prof. Spearman: Give each individual a rank-number according to his score for each of the two qualities to be compared, and find Σd^2, where d is the difference between his two rank-numbers. If the order of n individuals were shuffled at random, the average value of Σd^2 would be $n(n^2-1)/6$. The correlation coefficient is then given, roughly, by

$$1 - \Sigma d^2/\text{average } \Sigma d^2, \text{ i.e. } 1 - 6\Sigma d^2/\{n(n^2-1)\}.$$

Apply this rule to the figures of example 8.

12. Prove the statement in the above example that the average value of Σd^2 is $n(n^2-1)/6$. (Consider the order $1,2,3,\ldots,p\ldots,n$. The letter p might be shuffled into any one of n different positions. Counting gain of rank only, the average d^2 for the letter p is $\{1^2+2^2+\ldots+(p-1)^2\}/n$. This must be evaluated in terms of p and n, then summed from $p=1$ to $p=n$, and finally doubled, so as to include loss of rank as well as gain.)

13. Apply Spearman's rule to the orders of the English and Mathematics marks in Exercise 7, taking the orders as follows:

E	M	E	M
7	1	4	11
9	16	19	19
8	15	2	12
18	20	10	7
21	13	6	10
12	17	1	3
16	9	5	5
17	21	14	6
20	14	15	8
3	2	11	4
13	18		

ANSWERS TO EXAMPLES

EXAMPLES I

Page

4. **1.** (1) $a^{\frac{2}{3}}$,   (2) $(x-a)^{-\frac{1}{2}}$,   (3) a^2,

    (4) a^{-6},   (5) a,   (6) $a^{\frac{1}{3}}$.

2. (1) $x^{m-n/2}$,   (2) $(x-a)^{2m-n}$,   (3) $x^{13n/6}$,

    (4) x,   (5) $x^{(n-m)/n}$,   (6) x^n.

3. (1) $a^{\frac{3}{8}}$,   (2) $a^{\frac{9}{8}}$,   (3) $a^{-\frac{3}{8}}$,

    (4) $a^{\frac{2}{3}}$,   (5) a,   (6) $a^{2/(n+1)}$.

4. $x^{\frac{4}{3}}+2+9x^{-\frac{4}{3}}$, $18\frac{9}{16}$.   **5.** $v^{-\frac{2}{3}}$; $n=-\frac{2}{3}$.

6. $\frac{1}{5}$.   **7.** 1000, 25, 116.   **8.** a^{-4}.

9. (1) $\log \dfrac{a+x}{a-x}$,   (2) $\log (a+x)^5$,   (3) $\log \dfrac{1}{\sqrt[5]{(a+x)}}$,

    (4) $\log x^{2m+3n}$,   (5) $\log \dfrac{(a+x)^m}{(a-x)^n}$,   (6) $\log \dfrac{1}{y^k}$.

10. (1) b/a,   (2) $y^{1/n}$,   (3) ab^n,   (4) $a^{n/m}$.

11. (1) 2·322,   (2) 1·183.

5. **15.** (1) 2·3026,   (2) 2·079,   (3) $-0·223$.

16. 0·562.

17. (1) $\sqrt{a}/a$,   (2) $\sqrt{(x^2-a^2)}/(x-a)$,

    (3) $(\sqrt{x}+\sqrt{a})/(x-a)$.

18. (1) 0·707,   (2) 0·414,   (3) 5·828.

19. (1) 14,   (2) 2,   (3) 0, 8.

20. $\frac{1}{2}\log_e \{(1+y)/(1-y)\}$.

21. (1) $-2/x^3$,   (2) $1/(3x^{\frac{2}{3}})$,   (3) $-1/(2x^{\frac{3}{2}})$.

22. $(v/v_0)^{0·4}$.

23. When $t=0$, $y=ab$; as $t\to\infty$, $y\to a$. a=ultimate size, b=ratio of size at first observation to ultimate size. (c is a 'rapidity factor'.)

24. $y=20x^3$.   **25.** $tv^{0·4}=10^c$.

6. **26.** $[M^{\frac{1}{2}}L^{-\frac{1}{2}}T^{-1}]$, $[LT^{-1}]$.   **28.** 5.

29. $1/(2\sqrt{a})$.

EXAMPLES II

14. **1.** (1) $y=40/x$,   (2) $y=13-x$,   (3) $y=3x-2$,

    (4) $y=\frac{1}{2}x^2$,   (5) $y=\frac{1}{10}x^3$,   (6) $y=300/x^2$.

15. **2.** $P=36W+14$.   **3.** $5, -3$; 3.

4. $y=200/x^2$, 12·5.   **5.** $16\frac{2}{3}$.

6. 2·5.   **7.** 35 m.p.h.

8. (1) $y=1/(2x)$,   (2) $z=\frac{1}{24}x^3$; 33·1 per cent.

9. $I=9s/d^2$.   **10.** 104 min.

Page

16. **11.** (1) $X = 0.58l$,   (2) $P = 0.62W + 10$,

       (3) $Q = 1.13P + 44.7$,   (4) $Q = -0.67x + 46.5$,

       (5) $Y = 60/X + 10$,   (6) $X^2 Y = -2.5 + 18 X^2$,

       (7) $Y = 1.26 X^{10}$,   (8) $Q = 0.50l^{-0.8}$.

17. **12.** $y = 2 + 10/x$.   **13.** $c = 2$, $n = 0.67$.

     14. $a = 2.3$, $n = 0.5$.

     15. 1861–1870, $c = 12.4$, $m = 0.10$;

         1891–1900, $c = 10.8$, $m = 0.10$.

EXAMPLES III

20. **1.** (1) $a < \frac{9}{4}$,   (2) $a = \frac{9}{4}$,   (3) $a > \frac{9}{4}$.

   3. (1) $-bc/a^2$,   (2) $(b^4 - 4ab^2c + 2a^2c^2)/a^4$,

      (3) $(b^2 - 2ac)/c^2$,   (4) $(b^2 - 2ac)/ac$.

   4. (1) $cx^2 + bx + a = 0$,   (2) $a^3x^2 + (b^2 - 3abc) x + c^3 = 0$,

      (3) $a^4x^2 + (2a^2c^2 - ab^2c) x + c^4 = 0$,

      (4) $a^2cx^2 + (b^3 - 3abc) x + ac^2 = 0$.

21. **5.** (1) $4p^2 - 2q$,   (2) $x^2 - (4p^2 - 2q) x + q^2 = 0$.

   6. $\{(8b^3 - 6abc)^2 - 2a^3c^3\}/a^6$; $acx^2 + 2b (a+c) x + (a+c)^2 = 0$.

   7. $25x^2 + 234x + 676 = 0$.   **8.** $\frac{2}{3}$, $-\frac{1}{3}$.

  11. $3\frac{1}{8}$, $\frac{3}{4}$.   **12.** 6, -2.

22. **18.** $2\alpha - \beta$, $2\beta - \alpha$.

   24. $-b/a$, c/a, $-d/a$; $(b^2 - 2ac)/a^2$, $(c^2 - 2bd)/a^2$.

EXAMPLES IV

26. **1.** $(x-1)(x+2)(x-3)$.   **2.** -1.

   3. 2, 3, -8.   **4.** $(x+2)(x^2 + x + 3)$.

   5. -44.   **6.** $2a^4$.   **7.** It is.

   8. $(1+ax)(1-ax+a^2x^2)$, $(3x-y)(9x^2 + 3xy + y^2)$.

   9. The expression $= -(b-c)(c-a)(a-b)$.

  10. $(b-c)(c-a)(a-b)(a+b+c)$.

  11. $(b+c)(c+a)(a+b)$.

27. **12.** $(b+c)(c+a)(a+b)$.

  13. $\Sigma a (b-c)^3$, $\Sigma a (b+c)^2 - 4abc$, $\Sigma a^2 (b+c) + 2abc$.

  14. 0, $2\Sigma ab$.   **15.** $1 + 2 + 3 + \ldots + n$, $5 + 7 + 9 + \ldots + (2n+3)$.

  16. $x (x-1)(x-2) + 3x (x-1) + x + 1$.

  17. A, B can be chosen so that

       $A + B = p$, $-A\alpha - B\beta = q$;

       $A = (Bp+q)/(\beta - \alpha)$, $B = (\alpha p + q)/(\alpha - \beta)$.

  18. $n (n+1)(n+2)(n+3) - 6n (n+1)(n+2) + 7n (n+1) - n$.

  19. $\Sigma \dfrac{A}{(a-b)(a-c)}$, $\Sigma \dfrac{-A(b+c)}{(a-b)(a-c)}$, $\Sigma \dfrac{Abc}{(a-b)(a-c)}$.

Page

27. 20. (1) $\dfrac{1}{x+1}-\dfrac{1}{x+4}$,

(2) $\dfrac{\frac{3}{2}}{x+1}+\dfrac{\frac{5}{2}}{x+3}$,

(3) $1-\dfrac{\frac{3}{2}}{x+5}+\dfrac{\frac{1}{2}}{x+1}$,

(4) $\dfrac{3}{x}-\dfrac{3}{x-1}+\dfrac{3}{(x-1)^2}$,

(5) $\dfrac{-\frac{2}{3}}{x+1}+\dfrac{\frac{2}{3}x+\frac{11}{3}}{x^2-x+1}$,

(6) $1-\dfrac{\frac{8}{9}}{x+2}+\dfrac{\frac{8}{9}}{x-1}+\dfrac{\frac{1}{3}}{(x-1)^2}$.

21. (1) $\dfrac{1}{x-1}-\dfrac{1}{x-2}+\dfrac{2}{(x-2)^2}$,

(2) $2+\dfrac{\frac{2}{3}}{x-1}-\dfrac{\frac{2}{3}x+\frac{4}{3}}{x^2+x+1}$,

(3) $\dfrac{\frac{1}{4}}{x+1}+\dfrac{\frac{3}{4}}{x-1}-\dfrac{x+\frac{1}{2}}{x^2+1}$.

28. 24. $x+\dfrac{\frac{9}{4}}{x-2}+\dfrac{\frac{7}{4}}{x+2}$.

25. $x^2-3x+13-\dfrac{51\frac{2}{5}}{x+4}+\dfrac{\frac{2}{5}}{x-1}$.

26. $\log\{(x-3)^{\frac{11}{6}}(x+3)^{\frac{7}{6}}\}$.

27. $1-5x+21x^2-85x^3+\dots$; x must lie between $-\frac{1}{4}$ and $\frac{1}{4}$.

28. $ab=2$, $a^3+b^3=7$; $y^2-7y+8=0$; $-2\cdot90$.

EXAMPLES V

32. 1. 900.　　**2.** 259,974.　　**4.** 676.

5. 720.　　**6.** 120.　　**7.** 120.

8. 7,290,000, 729.　　**9.** 32,760, 6435.　　**10.** 105, 3.

33. 11. 705,432 ($=19.17.14.13.12$), 352,716.

12. 10, 60.　　**13.** 255.　　**14.** 8.

15. 43,046,721 ($=3^{16}$).　　**16.** 2,598,960 ($=52.51.49.20$).

17. 162,456,840 ($=30.29.28.27.19.13$).

18. 4095, 220, 298.　　**19.** 2730.

20. (1) 42,　(2) 21.　　**21.** 30.

22. (1) 220,　(2) 660.

34. 23. 12, 24.　　**24.** 3432 ($=_{14}C_7$).

25. 282,240.　　**26.** 362,880 ($=9!$), 60,480.

30. 8, 27;　(1) $\frac{1}{64}$,　(2) $\frac{1}{16}$,　(3) $\frac{1}{4}$,　(4) $\frac{27}{64}$.

35. 31. (1) $\frac{81}{256}$,　(2) $\frac{27}{256}$,　(3) $\frac{9}{256}$,　(4) $\frac{3}{256}$,　(5) $\frac{1}{256}$.

32. 7 : 2 : 7.　　**33.** $\frac{25}{64}, \frac{15}{64}, \frac{15}{64}, \frac{9}{64}$.

36. 34. (1) 120,　(2) 20,　(3) 60,　(4) 10,　(5) 10.

35. 15. (This is best done by straightforward counting.)

EXAMPLES VI

39. 1. 10th row is 1, 10, 45, 120, 210, 252, 210, 120, 45, 10, 1.

2. (1) $1+8x+28x^2+56x^3+70x^4+56x^5+28x^6+8x^7+x^8$,

(2) $1+15x+90x^2+270x^3+405x^4+243x^5$,

(3) $1-3x+\frac{15}{4}x^2+\frac{5}{2}x^3+\frac{15}{16}x^4-\frac{3}{16}x^5+\frac{1}{64}x^6$,

(4) $16x^4-96x^3y+216x^2y^2-216xy^3+81y^4$.

3. (1) 256,　　(2) 1024,　　(3) $\frac{1}{64}$,　　(4) 1.

LA　　　　　　　　　　　　　　　　　　7

Page

40. **4.** (1) $\dfrac{13-r}{r} \cdot 2x$,   (2) $\dfrac{26-2r}{r}$.

5. (1) 7th,   (2) 5th.

6. (1) $\frac{228}{25}$,   (2) 75,000,000.

7. (1) 16,   (2) 6435.

9. $1+0\cdot16+0\cdot0112+0\cdot000448$; $1\cdot172$.

10. $0\cdot8330$.

11. $1+20x+190x^2+1140x^3$; 21; 2^{20}; 3rd and 4th are each $\frac{190}{36}$.

13. $1140x^3$, $0\cdot00114$.   **15.** $y^7-7y^5+14y^3-7y$.

16. $A=-5$, $B=6$, $C=-1$.   **17.** £$1\cdot34$.

18. 3, $0\cdot238$.

EXAMPLES VII

45. **1.** $\frac{4}{13}$.   **2.** (i) $\frac{1}{3}$, (ii) $\frac{1}{5}$, (iii) $\frac{1}{15}$, (iv) $\frac{7}{15}$.

3. $\frac{1}{8}$, $\frac{1}{4}$.   **4.** $\frac{1}{4}$, $\frac{1}{2}$.

5. (i) $\frac{1}{36}$, (ii) $\frac{1}{18}$, (iii) $\frac{1}{12}$.   **6.** (i) $\frac{1}{40}$, (ii) $\frac{1}{1800}$.

46. **7.** $\frac{1}{5}$, $\frac{1}{9}$, $\frac{1}{120}$.   **8.** $\frac{5}{28}$, $\frac{125}{512}$.

9. $\frac{2}{21}$, $\frac{2}{81}$.   **10.** $(\frac{1}{4}+\frac{3}{4})^5$; (i) $\frac{45}{512}$, (ii) $\frac{53}{512}$.

11. $\frac{280}{2187}$, $\frac{11}{243}$.

12. (i) $\frac{1}{16}$, (ii) $\frac{15}{64}$, (iii) $\frac{15}{32}$.   **13.** $(\frac{1}{3}+\frac{1}{3}+\frac{1}{3})^5$, $\frac{10}{243}$.

47. **14.** $17:3$.   **15.** $\frac{32}{663}$.

16. $\frac{251}{180}$; the excess is his source of profit.

17. 6, £52; yes, he gains $\frac{1}{12}$ of the money staked.

18. $0\cdot214$, $0\cdot195$.

48. **19.** $\frac{2}{5}$, $\frac{11}{15}$.

20. $0\cdot0455$; (a) $0\cdot00048$, (b) $0\cdot0450$.

21. $0\cdot00262$.   **22.** $0\cdot813$, $0\cdot007$, $0\cdot00834 \ (=\frac{7}{839})$.

23. (1) $0\cdot613$, (2) $0\cdot0936 \ (=\frac{77}{823})$, (3) $0\cdot213 \ (=\frac{175}{823})$.

49. **24.** (1) $0\cdot00779 \left(=\dfrac{4,181}{536,664}\right)$,   (2) $0\cdot202 \left(=\dfrac{108,360}{536,664}\right)$,

(3) $0\cdot156 \left(=\dfrac{83,882}{536,664}\right)$,   (4) $0\cdot129 \left(=\dfrac{69,120}{536,664}\right)$.

25. $0\cdot107$.

EXAMPLES VIII

56. **1.** (1) $3n+2$; $(3n^2+7n)/2$,   (2) $3^{n-1}/2^{n-4}$; $16\{(\frac{3}{2})^n-1\}$,

(3) $3^n/4^{n-4}$; $768\{1-(\frac{3}{4})^n\}$,   (4) $7-3n$; $(11n-3n^2)/2$,

(5) n^2; $\frac{1}{6}(n+1)(2n+1)$,   (6) $2n^2+1$; $\frac{1}{3}(2n^3+3n^2+4n)$,

(7) n^2+n; $\frac{1}{3}n(n+1)(n+2)$.

2. 1240, $18\frac{2}{3}$.   **3.** 2, 1524.

4. (1) $a(2-n)+b(n-1)$; $\frac{1}{2}n\{a(3-n)+b(n-1)\}$,

(2) $a^{2-n}b^{n-1}$; $a^{2-n}(a^n-b^n)/(a-b)$.

ANSWERS99

Page

56. **5.** $(1-21x^{20}+20x^{21})/(1-x)^2.$

6. $\dfrac{a}{1-x}+dx\,\dfrac{1-x^{n-1}}{(1-x)^2}-\dfrac{\{a+(n-1)\,d\}\,x^n}{1-x},$

i.e. $\dfrac{a\,(1-x-x^n+x^{n+1})+d\,\{x-nx^n+(n-1)\,x^{n+1}\}}{(1-x)^2}.$

7. (1) $\tfrac12 n\,\{2a+(n+1)\,d\},$ (2) $(ab^{n+1}-a)/(b^{n+1}-b^n),$
(3) $n\,(n+1)\,(n+5)/3,$ (4) $\tfrac14\,(n^4+6n^3+7n^2+10n).$

8. $n^3+n-1;\ \tfrac14\,(n^4+2n^3+3n^2-2n).$

9. $\dfrac{-2}{1.2.3},\ \dfrac{-2}{2.3.4},\ \dfrac{-2}{3.4.5},\ \cdots,\ \dfrac{-2}{(n-1)\,n\,(n+1)};\ \dfrac{n^2+3n}{4\,(n+1)(n+2)}.$

57. **10.** $-\dfrac{1}{r\,(r+1)},\ -\dfrac{2}{r\,(r+1)\,(r+2)},\ -\dfrac{3}{r\,(r+1)\,(r+2)\,(r+3)};$

$1-\dfrac{1}{n+1},\ \dfrac12\left\{\dfrac{1}{1.2}-\dfrac{1}{(n+1)\,(n+2)}\right\},$

$\dfrac13\left\{\dfrac{1}{1.2.3}-\dfrac{1}{(n+1)\,(n+2)\,(n+3)}\right\}.$

11. 13·1 approx. **12.** After 12 terms. **13.** 11.
14. 47. **15.** $a+b,\ 3a+b$; A.P.
16. 8000. **17.** 2, 6, 18.
19. $2\tfrac14,\ 3,\ 3\tfrac34.$ **20.** £7550, £8. 10s.
21. 556 cm. **22.** $126\tfrac12$ sq. in.

58. **23.** 9; 4·92 sec. **24.** 84, $\tfrac16 n\,(n+1)\,(n+2).$
25. 12,670 c.c. approx. **26.** £573·5 approx.
27. £886 approx. **28.** £22·4 approx.

29. $£\dfrac{a\,(100+r)}{r}\left\{1-\left(\dfrac{100}{100+r}\right)^n\right\}.$

EXAMPLES IX

67. **1.** (1) 64, (2) $\tfrac49,$ (3) $\tfrac{7}{33},$
(4) $x/(1-x^2),$ if $-1<x<1,$
(5) $axy/(1-ax),$ if $-1<ax<1.$

3. (1) Convergent if $-1<x<1,$
(2) Convergent if a is numerically greater than 1,
(3) Divergent, (4) Convergent if $x<-1$ or $\geqslant 1.$

4. (1) $1-3x+6x^2-10x^3,$ if $-1<x<1,$
(2) $1+x-\tfrac12 x^2+\tfrac12 x^3,$ if $-\tfrac12<x<\tfrac12,$
(3) $1-\tfrac13 x^2+\tfrac29 x^4-\tfrac{14}{81}x^6,$ if $-1<x<1,$
(4) $\tfrac14-\tfrac14 x+\tfrac{3}{16}x^2-\tfrac18 x^3,$ if $-2<x<2,$
(5) $\sqrt3\,\{1-\tfrac13 x-\tfrac{1}{18}x^2-\tfrac{1}{54}x^3\},$ if $-\tfrac32<x<\tfrac32,$
(6) $\tfrac15+\tfrac{1}{25}x+\tfrac{1}{125}x^2+\tfrac{1}{625}x^3,$ if $-5<x<5.$

Page

68. **5.** $1 + x + 2x^2 + 3x^3 + 5x^4$; $1 - x + x^3 - x^4$.

    6. $1 \cdot 013$; third term $= -0 \cdot 00018$.

    7. (1) $1 \cdot 0149$,   (2) $0 \cdot 9941$,   (3) $4 \cdot 0207$,
       (4) $33 \cdot 6241$,   (5) $1 \cdot 1303$,   (6) $9 \cdot 8995$.

    9. $\frac{7}{15}$.   **10.** $1 + \frac{2}{3}x + \frac{5}{18}x^2$.

    11. $1 + \frac{1}{3}x - \frac{1}{9}x^2$.

    12. (1) $1 - x + x^2/2! - x^3/3! + \dots$,
       (2) $1 + x^2/2! + x^4/4! + \dots$,   (3) $x + x^3/3! + x^5/5! + \dots$.

    13. $1 - x^2 + \frac{1}{2}x^4$.   **14.** $1/(1+x)$.

    16. $\log_e 2$.   **17.** $2 \cdot 398$.

    19. 5th and 6th terms are both $\frac{625}{24}$ ($= 26 \cdot 04$).

69. **21.** $y + y^2/2 + y^3/3 + \dots$.

    22. $a = \frac{1}{2}$, $b = \frac{1}{10}$, $c = \frac{1}{120}$.   **23.** $-(n-4)!/(6n)$.

    24. $\frac{5}{2}$.   **26.** $3 - 5x + 9x^2 - 17x^3$; $-\frac{1}{2} < x < \frac{1}{2}$.

    27. $\frac{5}{2} + \frac{7}{4}x + \frac{11}{8}x^2 + \frac{19}{16}x^3 + \frac{35}{32}x^4$; $x^n \{1 + 3/2^{n+1}\}$.

    29. $\dfrac{\frac{5}{9}}{x-2} - \dfrac{\frac{5}{9}}{x+1} + \dfrac{\frac{1}{3}}{(x+1)^2}$.

70. **30.** $b = c = \frac{1}{2}a$.

    31. $2\{x + x^3/3 + x^5/5 + \dots\}$, provided $-1 < x < 1$;

      $2\left\{\dfrac{1}{2n+1} + \dfrac{1}{3} \cdot \dfrac{1}{(2n+1)^3} + \dfrac{1}{5} \cdot \dfrac{1}{(2n+1)^5} + \dots\right\}$,

      provided $n > 0$ or < -1.

    33. $2 \cdot 1972$.   **34.** $\sqrt{(3v/\pi h)}$; $x/2h$.   **36.** $1 \cdot 92$ in.

71. **38.** $0 \cdot 00022$.

    39. 1;   (1) $0 \cdot 0189$,   (2) $0 \cdot 0103$,   (3) $0 \cdot 9708$.

    42. $\epsilon/(e^{\epsilon/kT} - 1)$.

EXAMPLES X

88. **1.** $192 \cdot 9$, $52 \cdot 9$; Q.D. $= 40$, $\frac{2}{3}$ S.D. $= 35 \cdot 3$.

    2. 2507, 295; median $= 2545$; quartiles are 2365, 2685; Q.D. $= 160$.

89. **3.** (a) Mean $= 86 \cdot 7$; S.D. $= 8 \cdot 18$; Q.D. $= 5 \cdot 5$.
       (b) Mean $= 85 \cdot 6$; S.D. $= 7 \cdot 44$; Q.D. $= 4 \cdot 5$.

90. **4.** $y = 0 \cdot 407x - 21 \cdot 8$.   **5.** $Y = 0 \cdot 743X + 26 \cdot 3$.   **6.** $0 \cdot 19$.

91. **7.** LE, $0 \cdot 73$; LM, $0 \cdot 51$; EM, $0 \cdot 50$.   **8.** $0 \cdot 69$.

92. **9.** $-0 \cdot 28$.   **10.** $y = -1 \cdot 01x + 76 \cdot 5$; $x = -0 \cdot 877y + 75 \cdot 1$; $-0 \cdot 94$.

93. **11.** $0 \cdot 51$.   **13.** $0 \cdot 55$.

INDEX